Also by Chris Jennings

HIV/AIDS in South Africa

The Facts and The Fiction

The HIV/AIDS Biological Primer

(Or, everything you always wanted to know
about HIV/AIDS, but were afraid to ask)

HIV/AIDS

The Facts

and

The Fiction

Health Alert Communications

Disclaimer

Any statement made in this document is not unequivocal. Data may exist in other databases which countermand the data presented here. All data are subject to change as scientific and medical knowledge and technology advance. Any interpretation of this data set and/or the actions purportedly undertaken by any individuals and/or organizations are open to question and/or alternate interpretation.

ISBN-10: 0-936571-11-X
ISBN-13: 978-0-936571-11-9
Version 1.0

HAC

Health Alert Communications
P.O. Box 65
Hudson, NH 03051
(617) 497-4190
www.healthalert.net

Statement on Homosexuality:

Homosexuality is part of the Human Equation

Statement of Purpose:

The purpose of this work is to reconfigure the conceptual

paradigm of the HIV/AIDS epidemic, such that resource

allocations and health care interventions better serve the

populations at need -- both those with and without HIV/AIDS --

worldwide.

Gay people have suffered horribly from the HIV/AIDS epidemic.

It is not appropriate to lay blame on the afflicted.

HIV / AIDS
THE FACTS AND THE FICTION
Table of Contents

List of Tables

Abbreviations and Definitions

Words and terms **Bolded** in the text are listed in either
in *Abbreviations and Definitions* or in the *Glossary*.

AIDS	Acquired immunodeficiency syndrome
AIDS prodrome	*See Glossary*
Algorithm	Step-by-step procedure for diagnosing medical conditions
Amenorrhea	Cessation of menstruation
Amoebae	Plural of amoeba
Amoeba	A single-celled protozoan that causes amoebiasis
Amoebiasis	A disease caused by *Entamoeba histolytica*, an amoeba
Antigen	Substance or object that triggers the creation of antibodies
Antigenic Determinants	Antibody binding site on antigen molecule
Assay	Test for determining the amount of a particular constituent
Asymptomatic	Without symptoms
Atrophy	Wasting; reduction in size and/or degradation of organ/tissue
BBC	British Broadcasting Corporation
BLV	Bovine leukemia virus
Caprine	Relating to goats
CD4	T4-lymphocyte (a.k.a. T4-cell or helper cell)
CD4/CD8 ratio	Ratio of CD4 to CD8 lymphocytes
CD8	T8-lymphocyte (a.k.a. T8-cell or suppresser cell)
CDC	Centers for Disease Control
CEAV	Caprine encephalitis anemia virus
Cell-mediated immunity	Immune response involving T lymphocytes
Cellular immunity	Immune response involving T lymphocytes
Chagas disease	Disease caused by protozoan *Trypanosoma cruzi*
Classical AIDS	AIDS as it manifests in the United States and Europe
CNS	Central nervous system (the brain and spinal cord)
Codon	A triplet of amino acids (coding unit); part of a genetic code
Cohort	A group of individuals who share a characteristic
Cryptosporidiosis	Enteric disease caused by protozoan Cryptosporidium
CVID	*See Glossary*
Cytomegalovirus	Virus common to humans and deadly during HIV infection
Cytotoxic	Cell-killing
Dermatoses	Diseases of the skin
EIAV	Equine infectious anemia virus
ELISA	Enzyme-linked immunoassay

Endemic	Native to particular people, locality, region, or country
Endemicity	Adjective: quality of being endemic
Endometritis	*See Glossary*
Endonuclease	Enzyme that cleaves nucleotide chains
Entamoeba histolytica	The amoeba that causes amoebiasis
env	HIV gene encoding viral envelope glycoproteins
Enteric	Relating to the intestine
Enteropathic	Relating to the disease of the intestine
Epidemiology	Study of disease in populations
Epitope	Antibody binding site on antigen molecule
Equine	Relating to horses
Etiology	Cause of a disease or abnormal condition
Factor VIII	*See Glossary*
Francophone	A French-speaking person
gag	HIV gene encoding viral capsid proteins
gag proteins	HIV viral capsid proteins (p17, p24, and p55)
Genome	The genetic material of an organism
Genus	The category of biological classification ranking above species
HBV	Hepatitis B virus
Hepatitis B	A disease caused by the Hepatitis B virus (HBV)
HHV-8	Human herpes virus 8 (also know as KSHV)
HIV	Human immunodeficiency virus
HTLV-I	Human T-cell leukemia virus, type I
HTLV-II	Human T-cell leukemia virus, type II
HTLV-III	Human T-cell lymphotropic virus, type III (a.k.a. HIV-1)
HTLV-IV	Human T-cell lymphotropic virus, type IV (a.k.a. HIV-2)
Humeral immunity	Immune response involving antibodies
IF	Immunofluorescence assay
Immunoglobulin	Antibody
Integrase	HIV enzyme that inserts HIV's genes into a cells' normal DNA
IP	Immunoprecipitation assay
Incubation	Time of infection before opportunistic disease onset
IV	Intravenous
KS	Kaposi's sarcoma
KSHV	Kaposi's sarcoma herpes virus
LAV	Lymphadenopathy-associated virus (a.k.a. HIV)
Lentivirus	The *Genus* to which HIV belongs (one step above *Species*)
Lymphatic system	*See Glossary*
Lymphadenopathy	Abnormal enlargement of the lymph nodes

Lymphocyte	White blood cells that fight infection and disease
Lymphomas	Usually malignant tumors of lymphatic tissue
Lymphotropic	Selectively infecting lymphocytes
Macrophages	Mobile cells that engulf cells marked with viral antigens
Malnutrition	Inadequate or unbalanced intake or absorption of nutrients
Mitogen	A substance used to test immunological allergic response
MPMV	Mason-Pfizer monkey virus
Morphology	Form and structure of an organism
MVV	Maedi-visna virus
Mycobacteria	Bacteria having certain fungus-like qualities
Mycobacterium tuberculosis	Bacterium responsible for tuberculosis
NCI	National Cancer Institute (United States)
Nematode	Roundworm
Neurotropic	Selective attraction to/infection of nerve tissue
NIH	National Institutes of Health (United States)
NIV	National Institute of Virology (Republic of South Africa)
Notifiable diseases	Diseases reported government authorities as mandated by law
NYC bathhouses	*See Glossary*
Oncogenic	Relating to tumor formation
Papular	Having small solid, usually conical elevations of the skin
Pathogen	An organism that causes disease
Pathogenic	Adjective: disease-causing
PCP	*Pneumocystis carinii* pneumonia
Phylogeny	*See Glossary*
PID	Primary immunodeficiency disorder
Plasmodium falciparum	A protozoan parasite, the cause of malaria in humans
pol	HIV gene encoding viral enzymes
Positive predictive value	*See Glossary*
PPV	Positive predictive value
Presenting symptom(s)	The first symptom(s) to appear in a disease or syndrome
Prevalence	The proportion of individuals in a population having a disease
Primates	Monkeys, apes (and humans)
Prodrome	The premonitory (warning) symptoms of disease
Proliferative	Adjective: to increase in number, i.e., cancerous
Prophylaxis	Preventative therapy
Protozoa	Plural of protozoa
Protozoan	Singled-celled organism more complex than bacteria and fungi
Pruritic	Characterized by itching
Restriction Patterns	*See Glossary*

Retrovirus	*See Glossary*
Reverse transcriptase	Enzyme that "reverse-transcribes" viral RNA into viral DNA
RSA	Republic of South Africa
SAIDS	Simian AIDS
SRV	SAIDS-related virus
Sepsis	Sickness caused by bacteria or their toxins
Sequelae	Plural of sequela: aftereffects of disease, condition, or injury
Serological	Relating to serum
Seropositive	Positive serum reaction for presence of antibody
Serum	Fluid part of the blood
SIV	Simian immunodeficiency virus
Spectrophotometer	Instrument for measuring the change of color in an assay
Strongyloidiasis	A disease caused by *Strongyloides stercoralis*, a nematode
Strongyloides stercoralis	Nematode (roundworm) that causes strongyloidiasis
Syncytia	*See Glossary*
T-cell	T lymphocyte
T4-cell	T4 helper cells; targets for HIV infection (a.k.a. CD4)
T8-cell	T8 lymphocyte (a.k.a. T8-cell, suppresser cell and CD8)
Taxonomy	*See glossary*
Thymus (Thymic)	(Relating to) lymphatic system organ in which T-cells mature
Toxoplasmosis	Disease caused by a protozoan parasite *Toxoplasma gondii*
Toxoplasma gondii	Protozoan parasite, commonly infects CNS of AIDS patients
Tropism	Movement of an organism towards or away from stimuli
Ungulates	Animals with hooves; such as horses, goats, cows, and sheep
Viremic	The presence of viruses in the blood
WHO	World Health Organization
Zoonotic	Relating to disease transmission from animals to humans

Preface

In preparation for this book, **HIV/AIDS – The Facts and The Fiction**, and its adjuvant, **HIV/AIDS in South Africa – The Facts and The Fiction**, the Author reviewed more than 3,000 scientific and medical articles.

The scientific literature is not all-encompassing. Nor is it particularly methodical in its acquisition of knowledge. It is a collection of works by various authors having a myriad of differing approaches, disciplines, and agendas. Nevertheless, the scientific and medical literature represent, at best, the collective embodiment of scientific and medical knowledge, as well as the framework and justification for medical and/or societal interventions.

Given a sober review, the scientific literature is clear: (1) New York City is the epicenter of the AIDS epidemic; (2) the theory that HIV came from monkeys is a fallacy; and (3) the African AIDS epidemic-as-holocaust never manifested.

Unfortunately, a series of interlocking misconceptions have distorted scientific and public perceptions of HIV and the AIDS epidemic. In this distortion, the basic theorem for the origin-of-AIDS-in-Africa hypothesis is that HIV was **endemic** in Africa for 30 or 40 years before ecological and sociological changes forced it out of the jungle. Once exposed to naïve, urban, highly susceptible populations, the disease spread exponentially; eventually infecting tens or hundreds of millions of Africans (without anybody noticing) before reaching across the Atlantic to selectively infect gay men in New York City. This theory forms the foundation of scientific belief and, as such, has created a conceptual prism such that scientific and medical data are viewed as through a warped lens.

The goal of this work is to reconfigure the conceptual paradigm of the HIV/AIDS epidemic; such that resource allocations and health care interventions work to serve the benefit, and not the detriment, of the populations at need. As such, this work attempts to address all the relevant disciplines, present information previously overlooked, and reconfigure the AIDS scenario into a new conceptual constellation, the constellation depicted by the medical and scientific literature.

This book should not be considered the definitive work on HIV/AIDS — a project beyond the capability of any one person. Ideally, this book is a starting point for new investigations. Although the book contains many numbers, it should be viewed as a qualitative work and not a quantitative work. Its contents are derived principally from the scientific literature, but the scientific literature is not a perfect lens for viewing the world; nor is the Author a perfect medium for communicating the vastness of this collective knowledge.

For people unfamiliar with the concept, the method of grading in school systems can be quantitative or qualitative. A *quantitative* grading system assigns numerical grades to student scores. A *qualitative* grading system scores the outcome only as *Pass* or *Fail* (alternately as *Good, Fair,* or *Poor*, etc.). Therefore, after reading this book, one should be able to determine whether the current conceptual paradigm of the HIV/AIDS epidemic receives a passing or failing grade.

Sometimes the Author is viewed as seemingly uncaring in his discourse. The Author would like to apologize in advance should he hurt the feelings of anyone who has suffered a personal loss to AIDS. From the remote position as researcher and writer, the Author fears that he writes dispassionately, but hopefully, too, also from a position best to describe and characterize this massive event in human history, and in a fashion that might alter our perspective and interaction with it.

The use of certain words in this document requires some explanation. The following words are used to humanize the story and simplify its rendition. These words are not meant in any way to disparage:

Gay Men – Generally, the gay men described in this document were the subset who were exposed to HIV and/or developed AIDS early in the epidemic. The majority of these gay men were highly sexually active. Therefore, the term "gay men" or "homosexual males" in this document most frequently describes a subset of highly sexually active, gay men in specific geographic locales. This is not the behavior of all gay men. Not all gay men engage in high-risk sexual behavior. The term "gay men" is more colloquial than "homosexual men" or "men who have sex with men" and easier to use while writing. However, these terms are used by the Centers for Disease Control and Prevention (CDC) and most other public and private authorities discoursing on the topics and themes relating to HIV/AIDS (there being the legitimate concern that many men who have sex with both men and women do not self-identify themselves as gay).

Racial Terms – the terms used to describe race herein are the terms used in the medical literature of the times (and remain colloquial throughout much of the United States today): *white* designates Caucasian; *black* designates Negro/African-American born in the United States; *Hispanic* designates the Spanish speaking and/or relatively recent immigrants from Central and South America (and Puerto Rico) and/or their descendants.

AIDS and HIV Infection – the terms "AIDS" and "HIV infection" are used rather interchangeably throughout this document because the term "AIDS" was used to describe the new, mysterious disease for three years, give or take, before HIV was discovered. Therefore, the term "AIDS" was the descriptive term used in the scientific literature. After the development of a classification system, "AIDS" technically came to be defined as the final, terminal stage of HIV infection.

<u>The New York Times</u> – The Author does not intend to single out the *New York Times* for derision. The *New York Times* is used to represent the reportage of the national and international media because it is both blessed and cursed to have an excellent search engine.

PART I

New York City
The Epicenter of AIDS

PART I

New York City – The Epicenter of AIDS

At the birth of the AIDS epidemic, AIDS was a mysterious, uncharacterized disease of an unknown **etiology** (cause). No one knew the Human Immunodeficiency Virus (HIV) existed. No available laboratory tests could detect its presence. AIDS was diagnosed only after the manifestation of strange and virulent infections and tumors in previously healthy young patients, amidst patient populations historically unaffected by such maladies. The rapidity of disease onset and death was rampant and terrifying.

The sudden emergence of AIDS as a mysterious disease of unknown etiology triggered widespread and untoward fear of contagion, secular and parochial damnations of the afflicted, floundering scientific theories, strident cries of a refortified gay militancy, and riot police with rubber gloves. Few people now, who weren't adults at that time, could conceptualize the extent of fear and consternation of that time, magnifying the belief that anyone who had engaged in sex within the prior 10 years could be infected.

The process of discovery and characterization of this new mysterious disease occurred in the absence of any established procedures for diagnosis or mechanisms for epidemiological surveillance. Yet, looking back in time, as documented in the medical and scientific literature, the world's first legitimate AIDS cases occurred in New York City, and New York City was the epicenter from which HIV/AIDS spread throughout the United States and the world.

Chapter 1

The First Legitimate AIDS Clusters

An exhaustive review of the medical literature reveals that the first reasonable cases of

AIDS were documented in New York City (NYC). A *reasonable* case of AIDS has having two

vital characteristics. First, any reasonable case of AIDS should display the classical, early onset,

AIDS-related, opportunistic infections. Second, any reasonable case of AIDS should be part of a

cluster of AIDS cases. Reasonable AIDS cases do not exist in singular isolation (i.e., far

removed in time and geography from all other documented AIDS cases). AIDS is a consequence

of HIV infection. By their very nature, infections are transmissible; that is, they pass from one

individual to another. Therefore, transmissible infectious diseases manifest in clusters.

Looking back in time, it is simple to identify the first AIDS clusters as they appeared in

the medical literature. The first cases were reported by infectious disease specialists in urban

hospitals; essentially, the regional authorities: doctors to whom general practitioners would refer

patients with inexplicable and recalcitrant infections. These infectious disease specialists were

the first to observe and report on these clusters of inexplicably virulent, deadly infections. By

April 1981, data had been compiled on three AIDS clusters in the United States: two in New

York City, and one in Los Angeles.

The world's first legitimate (*reasonable*) AIDS cases were documented in New York

City in 1979. In this review of the medical literature, the first patient diagnosed with AIDS was

a 27-year old gay, male hospital guard who denied drug use.[1] It is somewhat ironic and against

the odds that he was black (African-American): the next 25 AIDS patients would be white

(Caucasian), and AIDS came to be perceived as a white, gay male disease. AIDS was unknown

at the time and no name existed for it; this patient was diagnosed with *Pneumocystis carinii*

pneumonia (PCP)[*] in April 1979. PCP is a classic, early onset, AIDS-related opportunistic

---

[*] *Pneumocystis carinii* has been renamed *Pneumocystis jiroveci*; the older term is used in this document

infection, alone accounting for roughly half of AIDS deaths (43% to 58%) during the first years of the epidemic.[1-7]

This particular patient died before 6-month follow-up. He was one of 13 New York City residents diagnosed with PCP between April 1979 and April 1981 – the first AIDS cluster. Demographic and medical data were available for only 11 of these patients, 8 of whom had died by the time of data collation.[1] **

The second AIDS cluster consisted of four gay men in NYC with severe anal ulcers caused by *Herpes simplex* viral infections. (Virulent untreatable viral infections – such as herpes and cytomegalovirus – were also classic, early manifestations of HIV infection). In these four homosexual men, the anal herpes infections were rampant, untreatable, and terminal. Three patients had died within a 12-month period. This cluster included a man diagnosed in July 1979, representing perhaps the world's second documented legitimate case of AIDS.[1]

Los Angeles contained the third AIDS cluster: four gay men hospitalized with *Pneumocystis carinii* pneumonia (PCP); two of whom died within a 3-month period.[2] In June 1981, the Centers for Disease Control (CDC) published a report on this cluster in the *Morbidity and Mortality Weekly Report* (*MMWR*).[3] Many health journalist subscribe to MMWR and echo its reports, so this *MMWR* report of June 1981 gave AIDS its first national media exposure, essentially that of a medical oddity. Consequently, people who were adults at the time believe that AIDS started in Los Angeles, even though the first Los Angeles case of PCP was diagnosed in February 1981 — nineteen months after the first PCP case was diagnosed in New York City.[2, 4] (Reports on the two NYC clusters would not be published in the medical literature for another 6 months later – December 1981.)[1, 4]

One month later, the CDC reported the fourth AIDS cluster: twenty-six gay men (twenty in NYC and six in California); all with the malignant skin cancer Kaposi's sarcoma. Plus . . . ten

---

** not all these patients had AIDS: 1 patient had terminal **AILD** (angioimmunoblastic lymphadenopathy with dysproteinemia); others had questionable clinical profiles *vis-à-vis* AIDS, and the majority were IV drug users whereas HIV had yet to enter the IV drug using population at large

additional PCP cases in California; split between San Francisco and Los Angeles.[5] Kaposi's sarcoma (KS) is also a classical, early onset, AIDS-related, opportunistic condition, alone accounting for roughly one-third of AIDS deaths (17% to 43%) during the first years of the epidemic.[6-11]

CDC reported the existence of this fourth cluster in the July 3, 1981 issue of *Morbidity and Mortality Weekly Report*.[5] The same day, the *New York Times* echoed the contents of the *MMWR* report in a 900-word article entitled "Rare Cancer Seen in 41 Homosexuals," the number of cases in the title reflecting an increased accumulation of new KS cases since the time the *MMWR* report had been composed.[12] (Journalists specializing in health and medicine received review copies of *MMWR* prior to the official publication date, enabling the media to release this "news" on the same day as the official government report.) The flood of AIDS . . . and the accompanying media cascade . . . had begun. Excepting for those few individuals appropriately positioned to be prescient, no one could have possibly conceived that these first AIDS cases were the initial sparks of a conflagration that would consume over a million lives over the next 30 years.

Importantly, these simultaneous reports of the Kaposi's sarcoma cohort by *MMWR* and the *New York Times* demonstrate how AIDS entered the domain of public awareness, and how the seeds of later AIDS misconceptions were distributed by the very first medical and media disseminations. The first seed of the African theory for the origin of AIDS was planted by these dual reports. *MMWR* stated that Kaposi's sarcoma was exceedingly rare in the United States, but that Kaposi's sarcoma ". . . accounts for up to 9 percent of all cancers in a belt across equatorial Africa, where it commonly affects children and young adults."[5, 12] This fact was echoed the same day by *New York Times* (and echoed again by *Time* magazine on December 21, 1981).[13]

In the face of this burgeoning epidemic and a desire for medical information, the usual peer-review process of medical publications was necessarily by-passed. Rather than relying on standard reporting by medical journals, which required a 6- to 12-month lead time, *Morbidity and Mortality Weekly Report*, the CDC publication, became the vehicle in the United States for disseminating reports on AIDS epidemiology, medical knowledge, and diagnostic techniques. These CDC reports were immediately echoed and amplified by the media.

Generally, these media reports essentially were accurate in themselves, but the amplification of a select group of data points resulted in distortion of the size, scope, and nature of the AIDS epidemic, the characteristics of AIDS as a disease, and the characteristics of the populations most affected by HIV infection. These distortions today continue to afflict both the medical community and the lay public alike. Moreover, AIDS was an eminent mystery begging a solution, and this first unfortunate association between KS and Africa helped to establish an abstract framework linking AIDS with Africa, setting the stage for later erroneous events.

These initial reports and subsequent reports quickly established that *Pneumocystis carinii* pneumonia (PCP) and Kaposi's sarcoma were AIDS-defining conditions. Initially, it seemed that PCP was the most common presenting symptom; it manifested alone in one-half of all AIDS patients.[6-10, 14-18] (A **presenting symptom** is the first symptom to appear in a disease or syndrome; frequently the symptom that prompts the patient to visit a doctor.) Kaposi's sarcoma presented alone in roughly one-third of AIDS patients.[6-8, 10, 15-18] Approximately six to seven percent of AIDS patients presented with both PCP and KS.

At first, oral candidiasis – the most common presenting symptom of HIV infection – passed virtually unnoticed because it did not cause mortality. Oral candidiasis, also known as oral thrush, eventually manifests in up to 80% of patients. It is caused by a fungal infection (*Candida albicans*).[19-23] (In the United States, oral candidiasis possibly, though rarely, occurs in absence of HIV infection.)

The relative incidence of these opportunistic infections in the AIDS patient population would change over time as effective prophylactic regimens and antiretroviral therapies came to the forefront.[24] However, in any untreated patient population suffering from AIDS, it is highly likely that a substantial portion of this population would manifest these AIDS-defining conditions (PCP, KS, and/or oral candidiasis) either alone or concomitantly – with the caveat that KS manifested principally among gay men.[25-27]

Table 1 – Early Onset Opportunistic Diseases in Classical AIDS

Name	Description
Pneumocystis carinii pneumonia (PCP)	Caused by fungus-like single-celled parasite, *Pneumocystis carinii*; common worldwide. Infects lungs. Previous to AIDS, found in kidney transplant patients whose immune system had been chemically suppressed. Occurs in 60% to 80% of AIDS patients. Initially responsible for 43% to 58% of deaths among AIDS patients, now brought under better control due to chemical prophylaxis; that is, chemically treating the patient before symptoms occur.
Kaposi's sarcoma (KS)	Malignant skin cancer. Appears first as pink, purple or brown lesions (wounds), usually on arms and/or legs; then spreading around body. In AIDS patients, may spread to gastrointestinal tract, lungs, and other internal organs. Initially occurred in 46% of homosexual AIDS patients, only 3.8% of heterosexual IV drug user AIDS patients; initially responsible for 17% to 43% of AIDS deaths. Onset was statistically associated in homosexual males with oral-anal sex and fecal contact possible. Kaposi's sarcoma herpes virus (KSHV) a.k.a. human herpes virus 8 (HHV-8) is now recognized as transmissible agent.
Candidiasis	Caused by species of *Candida*, a fungus common to skin, mouth, vagina, gastrointestinal tract of humans. In AIDS patients, usually takes oral form: white spots or patches on lateral sides of tongue, perhaps inside mouth on mucous membranes of cheeks; commonly lodges under nail beds and skin around armpits, groin, and rectum. Sometimes affects lungs. Frequently, first clinical (as seen in doctor's office) sign of HIV infection. Presenting symptom > 50% of cases.

Chapter 2

The First 1000 AIDS Cases in the United States

New York, Los Angeles, and San Francisco were quickly recognized as principal AIDS epicenters. Amidst the first 1,000 U.S. AIDS cases, the primary epicenters were: [11]

New York City	465 cases
San Francisco	121 cases
Los Angeles	69 cases
Miami	39 cases
Newark	27 cases
Chicago	18 cases
Houston	16 cases
Boston	12 cases

These 1,000 AIDS cases represented all AIDS cases reported to the CDC from June 1981 to February 1983. From the outset, NYC had prominence . . . a prominence persisting to present day. In other aspects, the characteristics of the U.S. AIDS epidemic would change dramatically over time; namely, the proportion of the high-risk groups relative to one another.

For example, by August 1981, only one risk factor for AIDS had been recognized: a history of homosexuality or bisexuality. At that time there were 108 AIDS cases: 107 males and 1 female. Of the males whose sexual preference was known, 95% were homosexual or bisexual (95/101).[6] In the first NYC cluster of PCP patients, seven patients had reported IV drug use as had one member of the Los Angeles cluster, yet IV drug users had yet to be recognized as a risk group.[3, 4]

By September 1982, most other high-risk groups had been recognized. Of 593 reported AIDS cases: [7]

Homosexual/Bisexual Men	73%
IV Drug Users	13%
Haitians	6%
No Known Risk Group	5%
Hemophiliacs	0.3%

These categories were hierarchal meaning that a person with two risk factors was placed into the larger of the two groups. The Haitian group excluded Haitians who were homosexuals and/or IV drug users. (Haitians were classified as a risk-group, temporarily.) The incidence of AIDS had doubled every 6 months since mid-1979, the exponential growth that would characterize the AIDS epidemic through the first decade. Sixty percent of those diagnosed over a year ago had died.[7]

Of these first 1000 AIDS cases, gay men accounted for 72% of the AIDS population with the IV population trailing far behind at 15%. Later in the epidemic, after sufficient inoculation of HIV into the IV population, the rate of growth among the IV population rapidly outstripped that of the gay population. (Needle transmission is a far more effective method of HIV transmission than any form of sexual congress.)[28] Afterwards, the relative percentages of IV drug users grew while that of gay males decreased, as shown in Table 2 (although gay males would continue to account for the majority of AIDS cases). These percentages represent cumulative AIDS cases to the respective date.[6, 14, 29-31]

Table 2 — The Relative Proportion of Risk Groups over Time

	1981	1983	1987	1993	2007
Gay/Bisexual	94 %	71 %	66 %	54 %	44 %
IV Drug Use	–	17 %	16 %	25 %	23 %
Both Gay & IV	–	–	8 %	7 %	7 %
Heterosexual	–	–	4 %	7 %	14 %
Blood or Tissue Recipient	–	–	2 %	2 %	1 %
Hemophiliac	–	0.6 %	1 %	1 %	~0 %
Undetermined	–	6.3 %	3 %	5 %	11 %
Haitian	–	4.5 %	n.a.	n.a.	n.a.

*categories were hierarchal – a person with 2 risk factors was placed into the larger group

This burgeoning IV population was centered in Newark, New Jersey, whereas the greatest burgeoning population of gay men with AIDS was next door in NYC. This geographic overlap of gay men and IV drug users in New York City and northern New Jersey, respectively, had been noted among this first 1,000 cases.[11] The IV population also subsumed blacks and Hispanics into the AIDS epidemic (the greater proportion of IV drug users with AIDS in this vicinity was black and Hispanic). IV drug use also accounted for the vast majority of heterosexual female AIDS patients, who were principally female lovers of male IV drug users; the majority of these women were black and Hispanic.

Once it was recognized that AIDS was transmitted by the use of shared IV needles, it should have been evident immediately that AIDS was a blood-borne disease. The transmission patterns of blood-borne disease and their associate risk-groups (gay/bisexual males, IV drug users, and health care workers exposed to blood and bodily fluids) had already long been established. Also, given blood-borne transmission, it could have been predicted that AIDS transmission would occur via blood transfusion and surgery. In addition, once AIDS was recognized as a blood-borne disease, it should have been evident that members of the general public outside these high-risk groups had a substantially lesser risk of contracting AIDS.

Although AIDS would eventually be acknowledged as a blood-borne disease, the actual pathological agent (HIV) had yet to be discovered. The etiology of AIDS remained a mystery. The disease was given its name, Acquired Immunodeficiency Syndrome, before its cause was known.

Chapter 3

AIDS Migrates Overseas

Early in the epidemic, the greatest risk factor for contracting AIDS was sexual contact with a gay man from the United States. This finding was consistent among incipient AIDS populations throughout Denmark, United Kingdom, France, West Germany, South Africa, and the Caribbean. A substantial portion of the first AIDS patients in all these locations were gay men who had homosexual contact with a man from the United States; such homosexual contact occurring either inside or outside the United States.

AIDS Migrates to Europe

In Europe, the majority of European AIDS patients were gay men, a few Americans among them. The European populations also contained a minority of black men; most whom were Francophones residing in French-speaking countries, or visitors from former French colonies. (**Francophones** are French-speaking people.)

Denmark

Perhaps the first incident of international AIDS transmission occurred between New York City and Denmark. In 1981, the world's first *reasonable* AIDS case occurring outside the United States was documented in Denmark: a 50-year old gay man who had visited **NYC bathhouses** annually since 1971. In January 1981, he presented with lymphadenopathy, fever, and weight loss. A lymph-node biopsy detected Kaposi's sarcoma shortly thereafter.[32-34] **Lymphadenopathy** means abnormal enlargement of the lymph nodes (glands).

The first 4 Danish AIDS patients all had homosexual contact either with residents of the United States or with a man who had visited the United States.[33]

In Denmark, at that time (1981), the greatest risk factor for AIDS was homosexual exposure to citizens of the United States in the years 1980 and 1981. This relationship was revealed years later by retrospective testing of stored blood samples.

The stored blood of 250 Danish gay men, collected in December 1981, had been tested. The blood of 22 men was found **seropositive** for the HIV antibody.[*] The travel history was known for 21 of these seropositive men. Eight (8) had visited New York City, San Francisco, or Los Angeles. Thirteen (13) seropositive men had never traveled to the United States, but 5 of these 13 reported sexual contact with American (U.S.) citizens outside of the United States.[34]

Seropositivity to HIV was significantly correlated with travel to the United States in 1980-1981 ($p < 0.007$). Seropositivity was also significantly correlated with a number of sexual partners in the United States ($p < 0.02$). In the words of the authors: "The results indicate that seropositivity first appeared in Denmark among Danish homosexual men who had visited the United States."[34]

This travel profile was confirmed by another study using a different immunological measure. Early in the epidemic, it was evident that **asymptomatic** gay men with inverted CD4/CD8 ratios were likely candidates for the yet unnamed AIDS. CD4 and CD8 are **lymphocytes** that stimulate and suppress **cell-mediated immunity**, respectively. CD4 is the **T-helper** cell, as known as the **T4-helper cell**. (The CD4 lymphocyte is also the host for HIV.) CD8 is the suppressor cell. In a healthy person, the CD4 cell population exceeds that of the CD8 cells; therefore, the CD4/CD8 ratio is greater than (>) 1.00. Once HIV infection has depleted the CD4 population, the ratio becomes *inverted*; that is, less than (<) 1.00.

Danish gay men who had traveled to the United States in 1980 and 1981 were 7.7 times more likely to have an inverted CD4/CD8 ratio compared with the non-traveling, gay, Danish population. As groups, both travelers and non-travelers had a similar number of sex partners per year.[35]

In surveys of two "promiscuous" populations of Danish homosexuals ($n \sim 80$), approximately 25% had visited the United States within three prior years, prompting one

---

[*] per an enzyme-linked immunoassay performed in a laboratory; commercial assays were not yet available

scientific author to write: "[The] high rate of travel of Danish homosexuals to the USA might explain the high rate of AIDS in Denmark which is at present the highest reported for citizens in a European country."[36]

West Germany

By December 1983, West Germany had a total of 44 AIDS cases, including one female. The male population was ninety percent gay – a mixture of European and American citizens residing in Europe. Fourteen of these patients had died. All 14 dead patients and all patients diagnosed before March 31, 1983 had a history of travel to New York, California, Florida, Haiti, or Central Africa – with one exception. (None of these men were Haitians or Africans.) The exception (the non-traveler) was the single hemophiliac who had died in 1982. Factor VIII concentrates used in West Germany are mainly of US origin.[37] (**Factor VIII**, a protein, is an essential clotting factor of blood plasma that is absent or inactive in hemophilia.)

In Munich, Frankfurt, and Berlin, the first emerging AIDS clusters had "direct contact with the US AIDS epidemic." West German health authorities anticipated an increase in the AIDS patient population paralleling that observed in the United States, but lagging 1.5 to 2 years behind.[37]

Compared with these 44 German AIDS cases, the United States had over 3000 reported AIDS cases by the end of December, 1983.[14]

United Kingdom

In December 1981, the United Kingdom reported its first AIDS case – a 49-year old gay male who presented with *Pneumocystis carinii* pneumonia (PCP). He traveled to Miami annually. His most recent visit had been 9 months before disease onset.[38]

A total of 14 AIDS cases had been reported to the Communicable Disease Surveillance Centre by July 31, 1983. Twelve patients were gay men and one reported IV drug use, as well. One patient was a hemophiliac whose Factor VIII had been imported from the United States.

The characteristics of the one remaining patient were not specified. Nine of these fourteen patients had visited the United States within 2 years before disease onset (lymphadenopathy). Among these 14 AIDS patients, two other members comprised a couple, and three members were casual sex partners with one another.[39, 40]

Clinicians in the United Kingdom anticipated an ". . . emergence of a problem similar to that in the US with a 2- to 3-year lag period."[39] Compared to these 14 AIDS cases in the United Kingdom at the time, their former colony, the United States, had 1,972 reported AIDS cases.[41]

The first report of an intravenous user with AIDS in the United Kingdom did not come until May 1985. By that time, the United States had 10,000 cumulative AIDS cases, with the IV population accounting for 17%. The first woman in the United Kingdom with (presumed) AIDS prodrome was reported about the same time.[42-44] (See *Glossary* for definition of **AIDS prodrome**.)

France

In July 1981, a 38-year old gay man presented with vague malaise. He developed PCP and oral candidiasis one month later: the first documented AIDS case in France.[45]

By early 1983, France had an emerging AIDS pattern that differed from patterns observed in other European and U.S. epicenters. The incipient AIDS populations in the United States, Denmark, Germany, and the United Kingdom has been almost exclusively gay, white males. The first French cluster contained a subset of self-reported heterosexual males and females, most of whom had Haitian or African citizenships. The self-reported characteristics of these AIDS patients in France (*n* = 29) are presented below in Table 3.[46]

Among this population, most of the admittedly gay men had visited the United States, notably New York, within the previous 5 years. One French heterosexual had undergone blood transfusion in Haiti. Four heterosexuals had lived in Equatorial Africa, including two native Zairians, but none of these four heterosexuals had visited the United States or Haiti.

By American standards, the population of females in the black cohorts was highly inflated. Whereas only one in the first 108 American AIDS patients had been a woman, the French cohort contained five women in twenty-nine people.[6, 46] It was not clearly established, at that time, that a transmissible agent was responsible for AIDS, or that women could readily contract AIDS via vaginal, as opposed to anal, intercourse or the sharing of IV needles.

Table 3 —The First Parisian AIDS Cluster

Gender & Sexuality		Nationality	
16	homosexual men	15	France
		1	Peru
8	heterosexual men	4	France
		2	Haiti
		1	Peru
		1	Portugal
5	women	3	France
		1	Zaire
		1	Haiti
Total		**29**	

In summary, five patients were black among these twenty-nine Parisian AIDS.* Also, four out of five non-homosexuals in this cohort had visited Central Africa, but had never been to Haiti or the United States. The cohort contained a large percentage of women, mostly black. By comparison, the United States had over 1200 AIDS cases at this time (March 1983), approximately 7% of them female.[14, 47] In addition, incipient AIDS bonfires had kindled throughout the urban centers of Europe, almost exclusively among white European (and American) gay men, some of them well-traveled. Yet given these meager data, members of the French Bureau of Epidemiology wrote: "We suggest that Equatorial Africa is an endemic zone for the supposed infectious agent(s) of this illness."[46] (HIV had not yet been discovered.) Other authors came up with the same theory for different reasons.[48, 49]

By the Fall of that year (October 1983), France had a total of ninety-four AIDS cases.

---

* assuming race corresponds with nationality, in that the first Parisian cluster contained 2 Zairians and 3 Haitians

Fifty-two of them were among admittedly gay men: 51 French men and 1 Haitian man. As with the other incipient European AIDS populations, this group of French AIDS patients had a history of travel: 57% had visited the United States; while 75% had visited the United States, Haiti, or Equatorial Africa.[50]

Surprisingly, roughly one-third of the Parisian population was African and Haitian nationals.[50] Also of note . . . half of the Africans were Zairian: nine out of eighteen. There were 12 women in the cohort: ten of them African or Haitian. Overall, only one Haitian man reported homosexuality. All other African and Haitian men ($n = 17$) reportedly denied homosexuality and IV drug use (with one man unknown). The Parisian cluster contained no IV users at this time.[50]

Table 4 — The Second Parisian AIDS Cluster

Nationality	Description (n = 94)
French ($n = 61$)	51 gay males 4 heterosexual males 2 heterosexual females 3 unknown males 1 hemophiliac
Africans ($n = 18$)*	10 heterosexual male 7 female heterosexuals 1 newborn
Haitians ($n = 10$)**	5 heterosexual males 3 heterosexual female 1 gay male 1 unknown male
Other ($n = 5$)	5

* Countries of origin (# patients): Zaire (9), Congo (4), Mali (2), Gabon (1), Cameroon (1)
** Countries of origin (# patients): England (1), Portugal (1), Spain (1), Peru (I), Pakistan (1)

Three years later, in 1986, a retrospective HIV antibody testing was performed on serum samples obtained from French hemophiliacs between 1981 and 1986.[51] Table 5 depicts the increasing prevalence of HIV antibodies among this hemophiliac population, a rate of growth similar to that observed among hemophiliacs in the United States, but "two years later than in the United States."[51]

Table 5 – Increasing prevalence of HIV antibodies among French hemophiliacs

Positive*	Year
0/8	1981
2/13	1982
6/33	1983
27/49	1984

*positive/number units tested

Hemophilia (Factor VIII)

Factor VIII is an essential blood clotting protein required by people with Hemophilia A. At this time, Factor VIII was a plasma-derived product, meaning the protein was extracted from the large production lots of pooled blood containing the blood of many people. Factor VIII products were capable of transmitting HIV before heat-treatment of the product became standardized (heat killing any contaminating HIV).

Among a cohort of Scottish hemophiliacs treated primarily with Factor VIII produced in Scotland, the rate of HIV seroprevalence was 15.6% ($n = 77$). By comparison, the HIV seroprevalence among a cohort of Danish hemophiliacs treated primarily with commercial concentrate imported from the United States was 59.1% ($n = 22$). This difference was statistically significant ($p<0.001$).[52]

In France, among a population of symptom-free, multi-transfused patients with hemophilias or hemoglobinopathies ($n = 425$), the highest rate of HIV seropositivity was "observed in hemophiliacs who received factor VIII concentrates prepared from plasma collected mainly on the American continent . . . ."[53]

As stated in a *Lancet* editorial: "The overall prevalence of AIDS in treated American hemophiliacs is about twice that in Europe, but in countries that use factor VIII concentrate from the USA the incidence is likely to increase."[54]

AIDS Migrates to Africa

Although the Republic of South Africa (**RSA**) is currently purported to be heavily afflicted with AIDS, early in the epidemic the emergence of AIDS among South Africans lagged far behind that of the United States.

In 1982, the first two documented AIDS cases in South Africa occurred in two gay men – both flight stewards who had visited the United States. Both died of *Pneumocystis carinii* pneumonia (PCP).[55]

Three years later (1985), serum samples from 843 South Africans, Kenyans, and Namibian were tested for HIV antibodies. By this time, there had been several published reports of AIDS among black Africans – principally the aforementioned Francophones – but also among Africans in Africa.[*] Several authors had also suggested that the yet-undetermined causative agent of AIDS might be endemic in Africa.[56-59] Scientists at South Africa's National Institute of Virology (**NIV**) had conducted seroprevalence studies with these concepts "in mind."[60] They used an **indirect immunofluorescence assay** in a laboratory; i.e., commercial HIV antibody tests were not yet available.

Out of 843 people tested for the HIV antibody, only 35 gay men in Johannesburg, South Africa tested positive. Twelve of these gay men had AIDS. The remaining twenty-three gay men had AIDS prodrome, defined as chronic lymphadenopathy syndrome. An additional 4 patients with apparent AIDS prodrome tested negative. (These tabulations are summarized in Table 6). The "majority" of these men had visited the United States or had homosexual contact with a man who had visited the United States.[60]

In comparison to these 35 South African, seropositive, gay men with AIDS or AIDS prodrome, the United States had a total of 10,000 AIDS cases at the time of this report (May 1985).[44]

---

[*] the Author will address these reports in the Chapter entitled "The First African Patients"

Table 6 – Populations tested in South Africa in 1985 (n = 843)

Test Result	Description	Race	Location
Negative	143 Adults	Black	Namibia/South West Africa
Negative	139 Adults	Black	Kenya
Negative	51 NIV staff	Black	Johannesburg, South Africa
Negative	67 Nurses	Black	Johannesburg, South Africa
Negative	96 Two-year olds	Black	Johannesburg, South Africa
Negative	75 Renal transplant patients	Black/White	Johannesburg, South Africa
Negative	50 Institutionalized children	White	Johannesburg, South Africa
Negative	49 NIV staff	White	Johannesburg, South Africa
Negative	76 Nurses	White	Johannesburg, South Africa
Negative	58 Lymphoma patients	White	Johannesburg, South Africa
Negative	61 Baboons	---	Kruger National Park, RSA
Negative	18 Vervet monkeys	---	Warmbaths, South Africa
Positive	*12 Gay men with AIDS*	*White*	*Johannesburg, South Africa*
Positive	*23 Gay men AIDS prodrome*	*White*	*Johannesburg, South Africa*
Negative	*4 Gay men AIDS prodrome*	*White*	*Johannesburg, South Africa*

By April 1987, sixty-four cases had been reported in South Africa. Fifty of these cases were among South African citizens; all of them being white males belonging to characteristic risk groups.[61] By the same report, the researchers had tested a group of "promiscuous black women" for the HIV antibody, promiscuous was defined as 56 prostitutes and 240 attendees of sexually-transmitted disease clinics.[61] One woman, an immigrant from Malawi, tested HIV seropositive.[61] [*] By comparison, at this time (April 1987), the United States had 33,952 reported AIDS cases.[62, 63]

The advent of AIDS in the Republic of South Africa (RSA) reflected a second pattern of international HIV transmission seen previously. In the RSA, Scotland, Denmark, and France, HIV was imported into the country via American blood products, namely plasma-derived clotting factors for the treatment of hemophilia.

In the RSA, 85% of hemophiliacs in the RSA treated with Factor VIII imported from the United States were seropositive for the HIV antibody (55/66) compared with 3% of hemophiliacs

---

[*] ELISA (Elavia, Pasteur Institute) and indirect immunofluoresence test (IF) with HIV-infected H9 cells

treated with local RSA product (3/98). (The American products were from large donor pools and the local products from small donor pools; on average, patients received similar amounts of Factor VIII per annum.)[64]

AIDS Migrates to the Caribbean

Haiti

The first publicized Haitian AIDS patients were residents of New York City and Miami. New York City had a cluster of 10 men and Miami had a cluster of 17 men and 2 women. The CDC reported them in July 1982.[65]

By September, Haitians accounted for a surprising six percent of the 593 AIDS cases in the United States. Most of these Haitians were men who had lived in the United States for less than two years.[7, 65, 66] Collectively, with a few exceptions, they denied homosexuality and/or IV drug use.[65, 67]

A substantial proportion of AIDS patients in these clusters presented with PCP, candidiasis, and/or other common AIDS-related conditions. However, Haitians both in the United States and Haiti exhibited more tuberculosis, disseminated tuberculosis, **CNS** (brain) **toxoplasmosis**, gastrointestinal conditions (diarrhea; **cryptosporidiosis**), weight loss, and candidiasis than did their white, gay, male American counterparts. [66, 68-70]

The initial Haitian AIDS populations — both in the U.S. and Haiti — contained more women than concurrent among the American and white European AIDS populations (no Africans had come to the forefront yet).[71, 72] Heterosexual AIDS transmission had not yet been documented. Therefore, it was suggested that AIDS was endemic among Haitians and/or Haiti, possibly due to a genetic component;[73] had been imported to the U.S. from Haiti;[66] and that perhaps voodoo practices were instrumental in its spread,[74-76] a thought echoed in the *New York Times*.[77]

Initially, only 15–18% of Haitians acknowledged homosexuality or IV use; whereas, 95%

of American AIDS patients could be attributed to these two risk groups. Eventually, reports came forward stating that the majority of AIDS patients in Haiti were bisexual, but these reports were not amplified in the media. By 1983, risk factors could be identified for the majority of AIDS patients in Haiti. According to several reports, 74–79% of AIDS patients in Haiti had acknowledged risk factors (bisexuality, blood transfusion, and/or multiple exposures to re-used hypodermic syringes). By one account, 72% of AIDS patients in Haiti had been identified as bisexual males who had at least one sexual encounter with visiting North-Americans or with Haitians residing in North America.[68, 72, 78, 79] Investigators had not been able to identify any AIDS patient in Haiti with a travel history to Africa; however, 10–15% of Haitian AIDS patients had visited the U.S. or Europe.[68, 72]

In the words of one leading Haitian researcher: "This increase [in the prevalence of gays] undoubtedly reflects our increased experience in obtaining sensitive information and a standardized approach, rather than any change in the epidemiology of AIDS in Haiti."[78] With this new information in hand, the CDC removed the category "Haitian" from its list of high-risk groups.[44, 72]

The epidemiological profile of the Haitian AIDS epidemic changed over time such that bisexuality decreased and heterosexual contact, including transmission between spouses, became viewed as the principal mode of transmission. Perinatal transmission also increased.[68, 72, 79] In the words of another leading Haitian researcher: "These data are consistent with hypothesis that HIV was introduced to the Caribbean by male homosexual tourists which may explain the high percentage of bisexuality found early on and the persistent, although decreasing, male predominance in patients with AIDS."[72]

Local investigative efforts sought to determine whether Kaposi's sarcoma existed in Haiti prior to 1978 or 1979, this being a possible marker to AIDS.[68] The investigators:

- questioned 21 practicing dermatologists and pathologists

- reviewed cancer biopsy records at three private Port-au-Prince hospitals (180 beds)

- reviewed 1000 cancer biopsies from hospital servicing 115,000 rural people

All investigations failed to detect the presence of any new KS cases prior to 1978–1979.[72] HIV antibody testing was performed on sera samples from 191 adults collected during a 1977–1978 dengue outbreak — all samples tested negative.[68] *

HAITI – A SNAPSHOT

The differentiation of AIDS from background morbidity and mortality in Haiti could be a challenge, given that, in the late 1970s, Haiti had an average life expectancy of 47.5 years, an infant mortality rate of 150 per 1000 live births, and an estimated daily intake of 1450 calories (a 35% daily deficit). Since 1969, in a community of 150,000 people, the most frequent conditions treated by the Hôpital Albert Schweitzer were, in descending order: nutritional deficiency diseases, pneumonia, diarrhea, malaria, and tuberculosis. The average annual income was $304 United States Dollars in 1980.[80]

Retrospectively viewed, the first plausible AIDS cases in Haiti occurred in 1980 or so (although the first case might have been an invalid outlier: CNS toxoplasmosis, an AIDS-defining condition but remotely possible in the absence of HIV infection). The emergence of the Haitian AIDS epidemic seemed virtually concurrent with the U.S.A. The Haitian epicenter was Port-au-Prince, a recognized holiday resort for American gay men and site of widespread homosexual prostitution.[68, 70-72, 78, 81-83]

The Other Caribbean

Both early and late in the AIDS epidemic, Haitians have been stigmatized as the possible harbingers of AIDS. In absolute numbers, in 1987, Haiti had the greatest number of AIDS cases in the world second to the United States. Yet AIDS statistics have a penchant for being reported in absolute numbers. When presented in an appropriate standardized format; i.e., *per capita*, the

---

* ELISA (whole virus), and radioimmunoprecipitation assay (RIPA) for p25 and gp120 HIV antigens

rank of Haiti falls to fifth place, trailing even the United States. Bermuda rises to the top as the country with the highest prevalence AIDS rate in the world, as seen in Table 7.[84]

Table 7 – AIDS Prevalence in United States and Caribbean in 1987

Country	Prevalence*	Reported AIDS cases	Population
Bermuda	76.3	42	55,000
Bahamas	30.5	68	223,000
United States	10.3	24,169	234,249,000
Trinidad/Tobago	9.4	108	1,149,000
Haiti	8.8	501	5,690,000
St. Lucia	8.4	10	119,000
Guadeloupe	3.5	11	315,000
Puerto Rico	2.5	81	3,179,000
St. Vincent	2.2	3	134,000
Martinique	1.9	6	308,000
Grenada	1.8	2	111,000
Barbados	1.6	4	251,000
Dominican Republic	1.0	62	6,248,000

* *per capita*, i.e., number of AIDS cases per 100,000 people

By one account published in 1988, Bermuda had an AIDS prevalence of 129/100,000.[85]

From another viewpoint, in 1986, the Caribbean nations with the most AIDS cases were among those with the strongest economic links with the United States; namely, the Bahamas, Haiti, Dominican Republic, Trinidad/Tobago, and Mexico.[86]

In several island nations, HIV seropositivity was correlated to sexual contact with North Americans.[86] Although the Haitian AIDS patients had categorically denied homosexuality, the greatest risk factor for HIV seropositivity in Trinidad, Jamaica, and the Dominican Republic was being a homosexual or bisexual man (according to seroprevalence data). For these Caribbean homosexual/bisexual AIDS patients, sexual contact with gay American men rather than promiscuity, per se, appeared to be associated with increased risk of infection.[72] In Puerto Rico, the AIDS epidemic seemed routed in IV drug use.[85]

Cuba, which remained embargoed by the United States, remained isolated from the AIDS epidemic, as evidenced by a seroprevalence rate 0.0147 (Cuba reportedly tested 1 million people and only 147 were seropositive).[72] Later surveillance studies would set the Cuban seropositive rate at 1/50,000, equivalent to 20 in a million.[87]

AUTHOR ANECDOTE

In 1981, during the summer that AIDS appeared, the Author had a conversation with a gay man in Provincetown, Massachusetts. It was a striking conversation whose significance the Author would not realize for years. This is an anecdotal account told second-hand. It would be designated as "hearsay" in a court of law and not eligible for presentation as evidence. As such, the reader should not grant this anecdote undue credence. It is a single data point from another universe, yet a graphic reality – an example of an enabling situation – by which AIDS might have been seeded into Haiti from New York City.

The man joyfully told the Author about his vacation in Haiti: "Club Med there is a big gay scene. And the gay magazines all have ads for sex clubs in Haiti. I bought an absolutely young beautiful man for the night. And the next day, he came back to visit me along with his wife and two children; all dressed in their Sunday best. And his wife came up to me, took my hand, and kissed it and said: *'Thank you for hiring my husband. Thank you. Thank you. Thank you. Thank you for hiring my husband!!'* All for $10 for the night! That's a third of the annual income of Haiti!" [*]

---

[*] this was an oral account; the 1980 per capita annual income was reported as $304 US by Barry M et al. Haiti and the Hôpital Albert Schweitzer. *Ann Intern Med.* Jun 1983;98(6):1018-1020

Chapter 4

The AIDS Epidemic Comes of Age

Globally, surveillance reports from the World Health Organization (WHO) support the old epidemiological tenet: "Disease follows trade routes." AIDS was undoubtedly an urban disease flourishing globally in urban epicenters, and these epicenters were linked by air to New York City. The spread of AIDS followed the tourist trade routes stemming from the eastern seaboard of the United States; therefore, AIDS appeared predominantly in Europe, the Caribbean, Brazil, and Canada. By December 1984, the primary, global, urban epicenters were New York City, San Francisco, Los Angeles, and Paris, as seen in Table 8.[88]

Table 8 – The Primary Global Urban Epicenters in December 1984

Primary Epicenters	AIDS cases	Prevalence*
New York City	3094	285.7
San Francisco	1055	254.7
Los Angeles	764	79.3
Paris	237	77.6
Geneva	10	62.0
Zurich	17	46.0
Amsterdam	28	41.4
Frankfurt	28	38.6
Copenhagen	26	45.5
Berlin	41	21.5
Madrid	15	4.7
Barcelona	8	4.5
Milan	4	4.0

* number of AIDS cases per million people; cumulative AIDS cases

Very likely, the city of Port-au-Prince, Haiti should be listed above. The AIDS prevalence rate in Haiti was 59.7, slightly below that of Geneva and Paris, and most of the initial AIDS cases manifested in Port-au-Prince.[88] However, no data set provided a direct comparison of Port-au-Prince to other global, urban epicenters.

In this statistical comparison provided by these authors, not all values were broken down to the city level. However, Table 9 provides statistical data another viewpoint: the countries with the most AIDS cases by December 1984:[88]

Table 9 – Countries with Highest AIDS Rate per capita in December 1984

Country	No. of AIDS cases	Prevalence*
United States	8297	35.9
Haiti	340	59.7
France	260	4.8
Brazil	182	1.4
Canada	165	6.6
West Germany	135	2.2
Belgium	65	6.6
Netherlands	42	2.9
Switzerland	41	6.3
Denmark	34	6.6
Australia	22	1.44
Spain	18	0.5
Sweden	16	1.9
Italy	14	0.3
Austria	13	1.7

* number of AIDS cases per million people

It is interesting that the seroprevalence of Haiti exceeds that of the United States early in the epidemic (1984), but by 1987 the seroprevalence of the United States outstrips that of Haiti, as shown previously in Table 7 (*allowing that* all Haitian AIDS cases reported here were legitimate).

There was apparently little or no AIDS behind the iron curtain. Both Poland and Yugoslavia reported no confirmed AIDS cases in 1985 while Czechoslovakia had two suspected cases (an African and his partner) that were later disqualified, failing to fit the CDC definition.[88, 89] The U.S.S.R. reported its first case in 1988.[90, 91] In 1990, an epidemiological investigation in Moscow found thirteen HIV-infected people.[92] Comparatively, the United States had approximately 132,000 AIDS cases at that time.[93]

Table 10 tabulates the most current figures for the U.S. urban epicenters. These numbers represent cumulative totals from June 1981 to year-end 2008.[94] New York City retains its predominance in absolute numbers.

Table 10 – US AIDS Cases by Metropolitan Statistical Area (2008)

City	Cumulative AIDS Cases[*]
New York City	214,870
Los Angeles	66,005
Miami	62,414
San Francisco	44,002
Washington, DC	36,328
Chicago	33,901
Philadelphia	32,402
Houston	28,563
Atlanta	26,404
San Juan, Puerto Rico	23,658
Baltimore	23,571
Dallas	22,220

No prevalence data are calculated from these figures, as such data would be meaningless. A vast proportion of the people represented by these numbers have passed away. Derivations of current prevalence estimates for HIV infection are available from Centers for Disease Control and Prevention as are surveillance reports of AIDS, this being a notifiable disease. **Notifiable diseases** are those required by law to be reported to government authorities

Also, direct comparisons of global epicenters have been rare. Such tabulations and comparisons of absolute numbers and seroprevalence are no longer featured in the medical literature, and such data have become increasingly obscure in reports circulated by international AIDS authorities.

---

[*] cumulative totals of AIDS cases from the 1981 to 2008 for the respective Metropolitan Statistical Areas (MSAs) For major cities, MSAs typically include neighboring cities and surrounding suburbs; for example, the NYC MSA includes Newark, New Jersey as well as suburbs in Pennsylvania.

PART II

The Monkey Fallacy

PART II

The Monkey Fallacy

In 1983, a team under Luc Montagnier of the Pasteur Institute, Paris, was first to isolate HIV. The virus was isolated from a white French male with lymphadenopathy; consequently, the virus was labeled **LAV** (lymphadenopathy-associated virus).[95] Soon thereafter, a team under Robert Gallo of the National Cancer Institute (**NCI**) isolated a virus from an AIDS patient.[96] This virus was labeled **HTLV-III** (human T-cell *leukemia* virus, type III) since the investigators thought it was closely related to HTLV-I, and -II.[97-99] Prior to the discovery of HTLV-III, HTLV-I and -II were the only two **retroviruses** known to reside in humans. Both were relatively rare leukemia viruses. All other known retroviruses infected only animals.

At first, HTLV-III was proclaimed related to HTLV-I and -II.[99] Then, as HTLV-III became differentiated from HTLV-I and -II, the meaning of the acronym "HTLV-III" changed from "human T-cell *leukemia* virus, type III" to "human T-cell *lymphotropic* virus, type III."[100-103] **Lymphotropic** means "attraction to lymphatic tissue and, in this application, to lymphocytes.

A bitter patent dispute ensued when Montagnier and Gallo determined LAV and HTLV-III had been the same virus. Gallo was essentially accused of the stealing the virus from Montagnier. The two institutions had shared resources, including a cell-line from the NCI that enabled HTLV-III propagation. Recently, a sober reviewer suggested that *contamination*, not theft, was the culprit that inadvertently introduced Montagnier's virus (LAV) into the cell cultures within the NCI.[104, 105]

Chapter 5

The Contaminated Monkey Theory

The theory that HIV came from African primates all started with a case of mistaken identity.

The African monkey theory began in November 1985, when a team of researchers at the Harvard School of Public Health isolated an "AIDS-like" virus from healthy, wild-caught, African green monkeys (*Cercopithecus aethiops*). This monkey virus was similar to HIV in that it was a **retrovirus** and it shared a number of biological traits with HTLV-III; namely, **morphology**, growth characteristics, and **T4-cell** tropism (*tropism*, in this setting, means selective infection of T4-cells). Moreover, the viral proteins of both the monkey virus and HTLV-III were similar in size and shared antigenic cross-reactivity (their antibodies reacted with each other's proteins).[106] * (An **antigen** is any substance or object that triggers the creation of antibodies.)

The monkeys were healthy. This "AIDS-like" virus did not cause any obvious disease or immunodeficiency. Nevertheless, these investigators labeled this virus **STLV-III$_{AGM}$** (*simian* T-cell lymphotropic virus, type III), obviously semantically mirroring the infamous HTLV-III. The subscript AGM designates African green monkey. (Prior to this discovery, monkey researchers had begun referring to a variety of different monkey diseases as "**SAIDS**," for simian AIDS.)

Several months later, the Harvard team isolated the exact same virus (STLV-III$_{AGM}$) from Senegalese prostitutes (presumably female, reports do not specify).[107] These healthy prostitutes were (1) free of disease, (2) free of immunodeficiency, and (3) had tested negative to HTLV-III antibodies. Given these circumstances, the investigators applied a second label to this exact same virus; it would be called STLV-III when found in monkeys, and **HTLV-IV** when found in humans.[107]

---

* cross-reactivity with *gag*- and 3' *orf*-encoded proteins; minimally with *env*-encoded proteins

Thus, given this nomenclature — plus the physical and biological characteristics of the virus — this "AIDS-like" virus was deemed the missing link between Africa and AIDS in the United States, seemingly identifying a likely African reservoir of retroviral progenitors for HIV. The *New York Times* published an article discussing the isolation of STLV-III$_{AGM}$ from Senegalese humans on March 27, 1984 — two weeks before the scientific report would appear in print.[108]

THE PURSUIT FOR MONKEY VIRUSES

Why was the Harvard research team mining for viruses in African green monkeys? They had learned of the purportedly heterosexual African cohorts: "A disproportionate number of AIDS cases has been reported in Central Africa; some of these cases were observed prior to recognition of the disease in the United States or Europe . . . We therefore investigated the possibility that primates were indigenous to Central Africa and were carriers of an infectious virus serologically related to HLTV-III/LAV,"[106] citing whom the Author calls the *Trilogy*. These are Clumeck *NEJM* 1984; Van de Perre *Lancet* 1984; and Piot *Lancet* 1984: the investigators who first reported on African clusters.[57-59] In the following years, hundreds and hundreds of journal articles would cite the Trilogy after any statement, to the effect: "AIDS is from Africa." Or such citations would trace back to the Trilogy as their source. The reports of the Trilogy are described in "The African Fallacy."

The actual findings of the scientific report were not exactly definitive. Eight of 50 Senegalese had tested seropositive for antibodies to STLV-III$_{AGM}$. A virus had been cultured from the peripheral blood lymphocytes of these eight people; a virus with "retroviral type particles, growth characteristics, and major viral proteins similar to those of the STLV-III and HTLV-III group of retroviruses."[107] Antibodies to this virus had "reacted strongly" with STLV-III **antigens** (viral proteins),* and "showed variable or no reactivity" with HTLV-III

---

* STLV-III antigens were gp160, gp120, p55, and p24 per radioimmunoprecipitation & sodium dodecyl sulfate-polyacrylamide gel electrophoresis

antigens. Nevertheless, the investigators wrote: "These results indicate that HTLV-III and STLV-III share common **epitopes** [antibody binding sites] in all the major viral antigens and that these are recognized bidirectionally across species lines . . . It is therefore conceivable that $STLV\text{-}III_{AGM}$ or HTLV-IV may have served as the progenitor virus to the human AIDS virus, HTLV-III/LAV; alternatively, they may have had a common progenitor."[107] The discovery of $STLV\text{-}III_{AGM}$ was instant scientific and media news worldwide. For example, this discovery was the focal point of an article "Linking AIDS to Africa Provokes Bitter Debate" in the *New York Times* on November 21, 1985.[109]

Then three years later -- the truth came out!

$STLV\text{-}III_{AGM}$ was not from Africa. $STLV\text{-}III_{AGM}$ was from the United States. The African samples — both the African green monkey and Senegalese prostitute samples — had been contaminated by a virus from the United States.

$STLV\text{-}III_{AGM}$ – the virus isolated from healthy African green monkeys –actually was isolated first from 4 sick Rhesus macaque monkeys, all living in the United States.[110] All four monkeys had an immunodeficiency disease characterized by some of the same early onset, opportunistic infections common to human AIDS patients. As a species, Rhesus macaques (*Macaca mulatta*) originate from Asia, but these particular four sick Rhesus macaques were residents of the New England Regional Primate Research Center (NERPRC) in Southborough, Massachusetts, one of several regional primate centers that bred primate populations for use in medical and biological experimentation.[110] Naturally, the virus derived from Rhesus macaque monkeys (MAC) was labeled **$STLV\text{-}III_{MAC}$**.

A team at NERPRC had discovered the $STLV\text{-}III_{MAC}$ virus and given samples to the researchers at Harvard School of Public Health.[111, 112] Evidently, the serum samples of the African green monkeys and human female prostitutes had become contaminated by $STLV\text{-}III_{MAC}$ in the laboratory at the United States.

Upon discovering STLV-III$_{MAC}$ in these contaminated samples, the Harvard team thought they had discovered a new "AIDS-like" virus, which they labeled STLV-III$_{AGM}$/HTLV-IV. In summary, the virus STLV-III$_{AGM}$ – the virus purportedly "discovered" in African green monkeys and Senegalese prostitutes, thought to be of African origin and proffered as the missing link between AIDS in humans and monkeys in Africa – was actually STLV-III$_{MAC}$: a homegrown American monkey virus isolated from sick Rhesus macaque monkeys living in the United States. **Therefore, the theory that AIDS originated in African monkeys arose from an incident of laboratory contamination**.

An excellent expository editorial entitled "A case of mistaken non-identity" in the February 18, 1988 issue of *Nature* summarized the lineage of events leading to the discovery.[111] The same issue also reported research demonstrating that STLV-III$_{AGM}$ and HTLV-IV were laboratory contaminants,[113] thereby invalidating their existence as a separate species.[111, 113, 114] [115-117] (Genetic sequences of the viral isolates in question – the **restriction patterns** – had been identical.)[113] Other researchers had suspected that STLV-III$_{MAC}$ might be masquerading as STLV-III$_{AGM}$/HTLV-4, but the issue was now clear.[118-120]

After unraveling the knotted puzzle, the author of the editorial added: "Such incidents of contamination are an enduring and confounding problem of laboratory work and not necessarily uncommon . . . This episode should serve as a strong warning for all virologists working with multiple isolates to check any new isolates against viruses present in the laboratory. I am aware, or have been told, of at least five instances in other laboratories in the United States and Europe where non-infected cell cultures became infected with HIV-1 in the same containment hood."[111] (If such publication were not enough, the author published another *Nature* editorial 4 months later entitled: "Human AIDS is not from Monkeys.")[114]

The author also commended the Harvard researchers for being forthright in acknowledging these incidents of contamination, stating: "Too seldom do researchers in this field retract data found to be erroneous."[111] This retraction from the head of the Harvard team

took the form of a letter published in the same February 18, 1988 issue of *Nature*.[121]

WHAT IS A RETRACTION WORTH?

After it was determined STLV-III$_{AGM}$ and HTLV-IV were laboratory

contaminants,[113] thereby invalidating their existence as separate species, the principal

investigator of the Harvard research team published a letter in the February 18, 1998

issue of *Nature* magazine, acknowledging this incidence of laboratory contamination.[121]

Eight months later, the October 1988 issue of *Scientific American* carried an

8-page article entitled "The Origin of the AIDS Virus," written by the same principal

investigator. This *Scientific American* featured a full-page photograph of an African green

monkey and failed to mention the incident of laboratory contamination. The short

recommended reading list contained an article published in April 1988, but failed to

include the retraction published in February 1988.[122]

Sadly, this finding of contamination seemingly passed unnoticed by the scientific

community at large. By the time the truth was published (February 1988), researchers had

discovered dozens of "AIDS-like" viruses in a variety of African primate species and

geographically-dispersed primate populations, all purported to have a viral relationship or

common ancestry with HIV. All were retroviruses (RNA-based viruses), like HIV. Perhaps by

force of habit, researchers labeled all these newly discovered monkey retroviruses as subspecies

of **SIV** (Simian Immunodeficiency Virus) although, again, infection by these viruses appeared

consistently benign, rarely inducing any observable disease symptoms in wild animals.[106, 114, 120,]

[123-125] To date, all pathogenic strains of simian viruses had been derived from captive

animals.[114, 117, 126-130] (In itself, this is not an important point; theoretically, pathogenicity could

result from cross-species transmission.)[131-133] Nonetheless, thereafter, whenever a related, truly

pathogenic monkey viral strain was discovered, researchers would instantly report this discovery, though generally such discoveries were singular in nature (i.e., in a single animal).[131, 132, 134, 135]

As the years of research ensued, successions of primate species were proffered as the original reservoir of either HIV or its progenitors. In the most recent rendition of this worn out story, the "Origin-of-AIDS" crown was bestowed to *Pan troglodytes troglodytes*, a species of chimpanzee.[136]

As far as the designations STLV-III$_{AGM}$ and HTLV-IV were concerned, other new viruses would step forward and assume these designations under the new **SIV** and **HIV** nomenclature, generally adopted in 1986 after recommendation by International Committee on Taxonomy of Viruses.[137, 138] All the various monkey retroviruses would be named one or another form of SIV; further obscuring pathogenic status and taxonomic relationships to the uninitiated.

Table 11 – The HIV Name Game

Acronym	Name	Now Known As
LAV	Lymphadenopathy-associated virus	HIV
HTLV-I	Human T-cell leukemia virus, type I	HTLV-I
HTLV-II	Human T-cell leukemia virus, type II	HTLV-II
HTLV-III	Human T-cell lymphotropic (leukemia) virus, type III	HIV
STLV–III$_{AGM}$	*Simian T-cell lymphotropic virus, type III — AGM*	*STLV–III$_{AGM}$*
HTLV-IV	*Human T-cell lymphotropic virus, type IV*	*HTLV-IV*
STLV–III$_{MAC}$	*Simian T-cell lymphotropic virus type III — MAC*	*SIV$_{MAC}$*
HIV	Human immunodeficiency virus	HIV
SIV	Simian Immunodeficiency Virus	SIV

Again, STLV-III$_{AGM}$, HTLV-IV, AND STLV-III$_{MAC}$ (all **Bolded** in Table 11, above) are *all* the same virus; correctly identified as STLV-III$_{MAC}$ — the laboratory contaminant innocently responsible for spawning the theory that AIDS came from African green monkeys. STLV-III$_{MAC}$

is a distinct, legitimate strain and is now known as SIV_{MAC}. All current viral players are now named HIV or SIV.

The name $STLV-III_{AGM}$ and the viral isolate designated as HTLV-IV by the Harvard team remain frozen through time in the scientific literature, distinguishing them from the legitimate viral isolates. However, other viral isolates designated as HTLV-IV have been renamed as HIV-2. The viral strain designated HIV-1 is recognized as the etiological cause of AIDS, and HIV-1 is the cause of the vast majority of legitimate AIDS cases throughout the United States, Europe, Asia, and Africa. HIV-1 was first theorized be have originated among African green monkeys, and later theorized to have originated from a rather rare species of chimpanzee (too be described in Chapter 10, *The Chimpanzee Has No Clothes*).

HIV-2 is a virus apparently endemic primarily in Western Africa; repeatedly reported as having a "close" relationship with HIV-1; purportedly originated in sooty mangabey monkeys; and purportedly inducing a form of immunodeficiency initially reported as AIDS; but later described as inducing a form of immunodeficiency less pathogenic than that caused by HIV-1. A sober discussion of the relationship between HIV-1 and HIV-2 – and an unraveling of the various related fallacies – must be reserved for another book.

AND THE JOKE IS . . .

The American public actually is widely infected by a virus that originally came from African green monkeys (AGMs) — but not HIV. The virus is SV-40 (simian virus 40) which was present in the kidney cells of African green monkeys. In the early 1950s, the kidney cells of African green monkeys were used to grow polio virus stocks for polio vaccines. These infected cultures contaminated the polio vaccine from 1954 to 1960. Millions of people worldwide were injected with these contaminated vaccines. SV-40 was also known as the *polyoma* virus in that it induced cancer in many animal models.[139-143] A recent Senate Hearing was held to evaluate its effects.[144]

Chapter 6

The Homegrown American Monkey Virus

As described previously, the virus thought to be from Africa (STLV-III$_{AGM}$/HTLV-IV) was actually STLV-III$_{MAC}$ – the homegrown American virus; the viral strain isolated from 4 sick Rhesus macaque monkeys living in Southborough, Massachusetts.

All 4 Rhesus macaque monkeys had an immunodeficiency disease with a "remarkable similarity" to human AIDS,[120] an immunodeficiency characterized by several opportunistic infections identical to those found in human AIDS patients (namely, candidiasis, cytomegalovirus, and cryptosporidiosis).[110]

Table 12 – Opportunistic Diseases of Monkey and Human

Name	Description
Candidiasis	Caused by species of *Candida*, a fungus common to skin, mouth, vagina, and gastrointestinal tract of humans. In AIDS patients, usually takes oral form: white spots or patches on lateral sides of tongue, perhaps inside mouth on mucous membranes of cheeks; commonly lodges under nail beds and skin around armpits, groin, and rectum. Sometimes affects lungs. Frequently, first clinical (as seen in doctor's office) sign of HIV infection.
Cytomegalovirus (CMV)	Normally present in salivary glands of humans and animals. Often widely scattered throughout the body in patients with advanced HIV infection. Causes problems in eyes, colon, lungs, liver, and adrenal glands. Suspected in promoting appearance of Kaposi's sarcoma. After PCP prophylaxis became effective, CMV infection became the major cause of mortality among AIDS patients. Cytomegalovirus is frequently spread in day-care centers, where it has been shown to survive on toys and Plexiglas for 30 minutes.
Cryptosporidiosis	An enteritis (inflammation/swelling of intestines) caused by *Cryptosporidia muris* and/or *C. difficile*; a one-celled parasite common to domestic and wild animals. Many minor, non-life-threatening outbreaks occur in day-care centers. In AIDS patients, may be major cause of mortality.

Initially, STLV-III$_{MAC}$ was the first and only STLV-III. It carried no subscript designation – referred to only as "STLV-III" – and was also a retrovirus.[110] When first isolated from the Rhesus macaques (and prior to all this confusion over mistaken identity), STLV-III$_{MAC}$

was described as morphologically "indistinguishable" from HIV.[110] The viral antigens of STLV-III$_{MAC}$ cross-reacted with viral antigens from HTLV-III (somewhat the equivalent of a "positive" result in the HIV antibody test).[145] * This cross-reactivity indicated that the viral proteins of both viruses were biochemically similar. (An **antigen** is any substance or object that triggers the creation of antibodies.)

STLV-III$_{MAC}$ also shared a number of other biological traits with HIV. The similarities between STLV-III$_{MAC}$ and HIV were observed via several experimental approaches, namely:

- *In vivo* (within the living systems), the biological activity of the STLV-III$_{MAC}$ strongly resembled HTLV-III in that both viruses selectively infected and killed the specific T-cell population that triggers the immune system into action against tumors, viruses and fungi.[110] **T-cells** are lymphocytes, a type of white blood cell. This cellular defect, the failure of this triggering mechanism, permits viral and fungal opportunistic infections and malignancies to grow unchecked.

- *In vitro* ("in glass;" i.e., in the laboratory), STLV-III$_{MAC}$ resembled HTLV-III in being **neurotropic** (able to infect brain cells) and induce AIDS-like lesions in brain tissue.[110] Moreover, STLV-III$_{MAC}$ induced the formation of syncytia. **Syncytia** are amorphous single-cell masses of cytoplasm containing multiple nuclei, cellular abnormalities formed by the fusion of multiple cells.[110] Syncytia formation is also a specific trait of HIV infection.[146-148]

In summary, STLV-III$_{MAC}$ ". . . has morphologic, growth, and antigenic properties indicating that it is related to HTLV-III/LAV" and, importantly, could transmit immunodeficiency.[149]

---

* radioimmunoprecipitation revealed STLV-III proteins of 160, 120, 55, 24 kilodaltans

The published report of this discovery was scientifically sober and remained within the scientific literary domain of virologists; never reaching general medical, media, or public awareness. In the scientific literature, the Rhesus macaque and STLV-III$_{MAC}$ were proposed as a nearly perfect analogous animal model of humans and HTLV-III; i.e., a living system for the testing of therapeutic agents and the development of vaccines.[110, 145, 149] Despite all these morphological and biological similarities between HTLV-III and STLV-III$_{MAC}$, no scientists presented any theories suggesting that STLV-III$_{MAC}$ was a relative or progenitor of HTLV-III, or at least not until STLV-III$_{MAC}$ was mistakenly thought to be a virus of African origin.

This erroneous African monkey origin-of-AIDS theory did not develop in a vacuum. First, a conceptual seed had been planted relating Kaposi's sarcoma to Africa. Second, the minority of European black men who denied homosexuality were being given undue weight and attention during this time. Third, faulty scientific methodology contributed to this fallacy; namely, the inappropriate use and interpretation of test results stemming from the first-generation diagnostic tests for HIV infection (as well as the use of unvalidated laboratory assays: "unvalidated" meaning there was no follow-up to ensure that laboratory findings actually translated into clinical disease over time).

Chapter 7

The Other American Monkey Viruses

Long before AIDS appeared in humans, monkeys in breeding populations at the California, Washington, and New England Regional Primate Research Centers had been plagued with immunodeficiency syndromes. In approximately 10 episodes between 1969 and 1984, various species of macaque monkeys were beset by several distinct constellations of opportunistic infections.[150] These episodes of disease were not called "immunodeficiency" at the time, but were retrospectively labeled "immunodeficiency syndromes" after the advent of AIDS. After AIDS appeared in humans, these macaque syndromes were collectively named *SAIDS* (simian AIDS), although the disease characteristics of SAIDS differed from species to species and virus to virus.

The first of these viruses, the Mason-Pfizer monkey virus (MPMV), was isolated in 1970.[151, 152] This retrovirus seemed to induce a true fatal immunodeficiency. Animals inoculated with this virus developed lymphadenopathy, **thymic** atrophy, and weight loss: and eventually died of opportunistic infections. (**Atrophy** is wasting; a reduction is size and degradation in composition of an organ or tissue.) However, immunodeficiency was not in vogue at the time, and MPMV disappointed its investigators by being unable to reliably transmit cancer: MPMV has been isolated from the breast tumor of a female Rhesus macaque, it being the early days of the Virus Cancer Program (back in the days when newly discovered viruses were likely to be labeled with terms like "**oncogenic** or "**proliferative**").[127-129]

From the late 1970s into the early 1980s, several regional primate centers were plagued by mysterious disease outbreaks presumably induced by viruses. It was generally assumed that viruses were responsible for these variant disease syndromes and occasionally viruses were isolated out of diseased monkeys, but no demonstrable viral causality could be found; that is, no virus proved consistent in being able to transmit disease when injected into healthy monkeys.

Finally, two viruses related to MPMV were isolated at the California and Oregon Regional Primate Research Centers, viruses that could reliably transmit one or another form of "immunodeficiency." AIDS was in the news at the time, so these viruses were named **SAIDS-related viruses** types I and II (**SRV-I** and **SRV-II**). The diseases they induced were characterized by the spontaneous development of **lymphomas** (malignant tumors) and **lymphoproliferative diseases**, and also a set of opportunistic diseases. SRV-II also induced **retroperitoneal fibromatosis**, tumors of fibrous or connective tissue lining the cavity of the abdomen.[130, 153-159] (Lymphomas and lymphoproliferative disorders are cancerous growths in lymph glands, the lymph glands being components of the immune system; thus, these disorders could be considered immunological disorders, though they contain no true T4-cell "defect," as seen in human AIDS.)

Collectively, these SRVs are called Type D retroviruses. They are not closely related to HIV and do not belong to the same genus as HIV.[127, 160, 161] (**Genus** is the biological classification one rank up from species, not that taxonomic disciplines necessarily apply to retroviral nomenclature.) In an experimental context, a successful vaccine was developed against SRV-1.[162] News of this success was circulated by the Associated Press and appeared in the *New York Times* on September 3, 1986,[163] prompting unrealistic expectations, as the challenge with HIV and human AIDS is far different.

The story at the New England Regional Primate Center in Southborough, Massachusetts, though overtly similar, was slightly different. Since approximately 1979, researchers had noted an increase in mortality due to what would later be called opportunistic infections.[150] The cause remained elusive until 1985, when STLV-III$_{MAC}$ was isolated from the 4 sick Rhesus macaque monkeys. The characteristics of the immunodeficiency induced by STLV-III$_{MAC}$ (the contaminant) were described as analogous to HIV infection in humans; i.e., a true immunological defect (characterized by depletion of the T4-cell population), unlike all other forms of primate "immunodeficiency" conditions proffered as simian AIDS.[110, 126]

Chapter 8

HIV's American Relatives

The prevailing notion is that HIV represents some new kind of virus; something like HIV had never been seen prior to AIDS. Genomically (the **genome** is the genetic material of an organism), HIV is an RNA-based virus. RNA stands for ribonucleic acid. Viruses based on RNA that replicate using a DNA intermediate are called **retroviruses** (see *Glossary*). Prior to HIV, most known viruses were based on deoxyribonucleic acid (DNA). HIV was only the third known human retrovirus. Although retroviral infections were rare amongst humans, a number of **pathogenic** (disease-causing) retroviruses were familiar to veterinarians and research virologists. Retroviruses commonly infect a number of domestic animals called ungulates. **Ungulates** are "hooved" animals. Those known to be susceptible to pathogenic retroviruses included goats, horses, sheep, and cattle, all serving as natural reservoirs for pathogenic retroviruses in the United States.

Domestic ungulate retroviruses included the equine infectious anemia virus (EIAV), the caprine arthritis-encephalitis virus (CAEV); and the ovine maedi-visna virus (MVV), also known as the visna virus. Equine means horse, caprine means goat, and ovine means sheep. All these ungulate retroviruses shared a number of characteristics with HIV, including morphology, antigenic reactivity, and/or the manner of growth and disease induction within the host. Taxonomically, all these ungulate retroviruses belong to the *Retrovirus* Family and the *Lentivirus* Genus, meaning they are all lentiviruses: they are slowly progressive; induce syncytia formation and cell death; infect cells of the immune system, although the specific target cell may differ (lymphocytes, monocytes, macrophages); and are neurotropic, as is HIV.[164]

When HIV was first discovered, a number of various studies identified HIV as a member of the lentivirus group, and attempts were made to determine the **phylogenic** relationship of HIV among the current set of known and familiar retroviruses, including HTLV-I, HTLV-II, and these ungulate retroviruses. (See *Glossary* definition of **Phylogeny**.) Categorically, the researchers agreed that HIV belonged to the lentivirus family and that HIV shared a common progenitor with all these retroviruses, and that HIV was phylogenically closer to the ungulate viruses than to HTLV-I and -II.[101, 103, 164-167] Typically, HIV was determined to have, **phylogenically**, the closest relationship to the visna virus of *sheep*, as determined by its infectious qualities, nucleoside sequencing, and/or amino acid identity (meaning codon preferences: a **codon** is a triplet of amino acids that comprise a coding unit; a unit of genetic code that starts or stops protein synthesis).[100, 102, 103, 164-166, 168-171] The visna virus is also known for developing antigenic variants (during replication) that can escape temporarily from host immune surveillance.[101] HIV is also characterized by this high degree of variability in the "envelope" proteins of its outer protein coat, a phenomenon known as "antigenic drift," a characteristic of lentiviruses that enabled one researcher in 1985 to predict that "development of a vaccine to prevent AIDS may prove to be a considerable challenge."[102] Table 13 lists several known lentiviruses.[164, 172, 173]

Table 13 – Lentiviruses Domestic to the United States

Lentivirus	Acronym	Host	Cell Tropism
Maedi-visna virus	MVV	Sheep	Macrophages
Caprine arthritis-encephalitis virus	CAEV	Goat	Macrophages
Equine infectious anemia virus	EIAV	Horse	Macrophages
Bovine immune deficiency virus	BIV	Cow	Macrophages
Feline immunodeficiency virus	FIV	Cat	T cells
Simian immunodeficiency virus	SIV	Primate	T cells
Human immunodeficiency virus	HIV	Human	T cells

Unlike the researchers of monkey viruses, these ungulate virus researchers did not attempt to link domestic ungulates to the origin of HIV; rather, they simply offered these various animal models of HIV infection as living laboratories for antiviral drug and vaccine development.

The other lentiviruses introduced in this book include HIV, STLV-III$_{MAC}$, and the many "AIDS-related" viruses derived from African primates. Still more lentiviruses have been isolated from mice, rats, house cats, lions, pigs, and birds.[106, 127, 158, 164, 174]

Chapter 9

The Monkey Fallacy Propagates

"Although these non-human primate lentiviruses are called immunodeficiency viruses, they do not induce an AIDS-like disease in their natural hosts."[125]

By February 1988, when the theory that "HIV came from monkeys" was discredited — by the discovery that STLV-III$_{AGM}$/HTLV-IV was actually the "American" virus STLV-III$_{MAC}$ — scientists had derived dozens of "immunodeficiency" viral isolates from a handful of African monkey species. The concept had gathered too much momentum. [Also, by that time, a French research team had isolated a legitimate HTLV-IV* from Senegalese humans, which stepped in to fill the slot vacated by the discredited STLV-III$_{AGM}$/HTLV-IV.[175, 176]]

Researchers in the US and Europe would continue isolating and characterizing the genomic structure of "immunodeficiency" viruses from African monkeys, comparing them with HIV, and declaring some degree of taxonomic relationship — all true to some remote extent: they are all **lentiviruses**, a genus of retroviruses, and share certain genetic and proteinaceous traits. (**Proteinaceous** means "relating to, resembling, or being protein.")

By 2002, scientists had reported SIVs in at least 33 African primate species, as determined by serological and/or molecular evidence. (**Serological** means the testing of serum for antibodies, **serum** being the liquid portion of blood.) It is interesting that, at this point in time, only African primates were found to be host to SIVs, and that these SIV infections were typically benign in their natural hosts.[125] [The only reliably pathogenic viruses that induced AIDS-like symptoms in non-human primates had been isolated from macaque species in captivity, namely the SIV$_{MAC}$ strains (the new name for STLV-III$_{MAC}$).] To repeat, macaques are from Asia: ". . . but macaques in the wild in Asia appeared not to be infected with SIV."[123, 125] In one report, none of 798 monkeys of varying species (all Genus *Macaca*) tested seropositive

---

* now called HIV-2, purported to be "closely related" to HIV-1

for SIV antibodies; the locales being from India, China, Taiwan, Malaysia, Indonesia, India, Sri Lanka, Malaysia, Philippines, Japan, and Thailand.[123]

The processes used for the isolation of viruses seem somewhat arbitrary. First, scientists must have a reason to investigate any particular potential host. Second, different viral "mining" techniques evidently can derive different viruses from the same host tissues. For example, after STLV-III$_{AGM}$ was first "discovered" in African green monkeys, a Japanese team attempted to replicate these findings — unsuccessfully. However, with a change of procedure, they were able to isolate 19 viruses from the same tissues.[111]

Nevertheless, it's somewhat ironic that the only two viruses widely recognized to transmit immunodeficiency consequent to a T4-cell defect (HIV and STLV-III$_{MAC}$) are both from the United States.

By the time STLV-III$_{MAC}$/HTLV-III was discredited, several notable luminaries had tried to dampen the flames on the African theory of AIDS origin, but such comments seemed relegated to relatively obscure publications such as *Skin & Allergy News* (a trade publication, not a journal). In the January 1988 issue of *Skin & Allergy News*, the following experts were quoted at an international AIDS conference: [177]

- **Dr. Luc Montagnier, Pasteur Institute, Paris:** The origin of AIDS viruses is "a continuing mystery . . . There is no evidence of any reservoir species of monkeys that is truly positive for HIV, even though some monkey viruses may resemble HIV more than they resemble the simian immunodeficiency virus . . . Some very weak arguments are used to place the origin of HIV in Africa." [95]   (Dr. Montagnier was the first to isolate HIV.)

- **Dr. Jonathan Mann, Director AIDS Program, World Health Organization:** "There are absolutely no data to support any hypothesis [about the origin of AIDS]."[177]

- **Dr. Peter Piot, Institute of Tropical Medicine, Antwerp, Belgium:** "The biologic gap between the monkey virus and human host is too wide to be bridged in a single step, even with direct injection of blood." [177]

This last comment is a rather surprising, coming as it does from a man whose publications formed a foundation for the African holocaust belief structure. (Also, years later, after he had become the Executive Director of UNAIDS, he was quoted by the *New York Times* as saying: ". . . the virus probably moved out of Zaire's rural areas to the cities, and then spread from the continent."[178])

Irregardless, opinions such as these were not widely echoed in the scientific and general media, and the clarion call of HIV/AIDS being related to Africa had been taken up by other players. Therefore, we currently have an obstructive nomenclature and taxonomic muddle regarding the ancestral relationships of the SIVs, HIVs, and other lentiviruses. Science and medicine, as with any other human endeavor, are influenced by fads and hearsay. Various belief structures persist long after they have been invalidated.

Thus, the hunt for the origin of AIDS, via presumed SIV progenitors, continues until today in *Myopia*, Africa.

Chapter 10

The Chimpanzee Has No Clothes

As described in Chapter 5 (*The Contaminated Monkey Theory*), an incident of laboratory contamination in 1985 spawned the theory that HIV originated in African primates. Although the discovery of this laboratory contamination invalidated this theory, this discovery escaped the notice of most scientists and the public at large.

In 1999, researchers would revise this theory of African primate origin by identifying a species of chimpanzee (*Pan troglodytes troglodytes*) as the original natural reservoir of HIV's progenitor.[136] This chimpanzee species is now accepted as the original reservoir of HIV in the scientific world.

Prior to this finding, three other strains of chimpanzee "immunodeficiency" viruses had been isolated and characterized.[179-183] However, only this new strain was "closely" related with HIV. This chimpanzee virus was isolated from a single chimpanzee (Marilyn), who was caught wild in Africa but exported at four years old to a primate facility in the United States, where she was raised as a breeding animal.[184]

The results of this genetic comparison, though widely disseminated throughout both scientific and public media, are highly suspect for a variety of reasons. Foremost, a single gene (*pol*) was used for this comparison. The gene, purportedly derived from Marilyn, was highly manipulated: four separate genomic fragments were amplified and reassembled into a single, allegedly resurrected, gene.[136]

At outset, Marilyn seemed a good prospect. In 1985, only she, of ninety-eight chimpanzees, had antibodies "strongly reactive" with HIV in five separate antibody assays. She had not been used in AIDS research, nor had she received any human blood products since at least 1969.[136, 184] She had died young, at twenty-six years old, a week following the birth of still-born twins. She died from pneumonia and toxic sequelae to the stillbirths. (**Sequelae** is the plural form of *sequela*: the aftereffect of disease, condition, or injury.) Autopsy revealed **endometritis**, retained placental elements, and **sepsis**).[136, 184] None of these autopsy findings were particularly strong indications of an HIV-type immune disorder. (Chimpanzees can be infected with HIV. They develop antibodies, but get a flu-like illness: they *don't* develop AIDS.)[185-189] Researchers were not able to isolate any live virus from Marilyn's autopsy tissues (lymph nodes, spleen, and brain).[184]

Fourteen years later (1999), a team of researchers extracted 4 genomic fragments from Marilyn's frozen spleen and lymph-node tissues. These overlapping genomic fragments were amplified and sequenced into a "complete proviral genome" which the investigators identified as its *own unique species*. This reconstructed virus was designated $SIV_{CPZ}US$ — the "US" meaning the United States.[136] By this point in time, three other chimpanzee SIV isolates now had been described:

$SIV_{CPZ-GAB-1}$	This virus was isolated from a chimpanzee wild-caught in **Gabon**, but had resided in captivity thereafter. Fifty wild-caught chimpanzees residing in captivity had been tested for HIV and two had tested positive; this chimpanzee was one of the two that tested positive [179, 181] *
$SIV_{CPZ-GAB-2}$	This 'virus' was reconstructed from genomic fragments isolated from the second chimpanzee that tested positive for HIV (fifty wild-caught chimpanzees residing in captivity had been tested for HIV and two had tested positive); these viral fragments were amplified and reconstructed to create this viral genome [182]
$SIV_{CPZ-ANT}$	This viral genome was created by amplifying and reconstructing six overlapping genomic fragments into a viral genome; the six fragments had been retrieved from a wild-caught chimpanzee that had been exported to Belgium, where it resided when this evaluation was performed [180, 183]

This reconstructed *pol* gene of $SIV_{CPZ}US$ was compared phylogenically with known HIV and these three chimpanzee SIV strains. The investigators concluded: ". . . [$SIV_{CPZ}US$] clustered well within this group, but was not particularly closely related to any one human or chimpanzee virus."[136]

---

* 2/50 Gabon chimpanzees tested seropositve for HIV-1; all chimpanzees were wild-caught but had resided in captivity thereafter

Yet, overall, the investigators concluded that $SIV_{CPZ}US$ clustered closer to the three HIV-1 groups (M, N and O) than with any of the other chimpanzee strains derived from *Pan troglodytes troglodytes*. Of the three HIV groups, Group M (major) is the overwhelmingly predominate viral group; it is responsible for HIV infections throughout the United States, the Caribbean, the Americas, Europe, and Central Africa (meaning Zaire, primarily). Group O (outlier) is self-descriptive, and Group N (new) had been documented only in two individuals in Cameroon at that time.[136]

Overall, these findings mean that the three HIV-1 strains "clustered closely" only with the SIV_{CPZ} strains derived from Marilyn ($SIV_{CPZ}US$) and the two Gabon chimpanzees (designated with the subscript GAB).[136] The fourth viral strain (subscript ANT) had been derived from another species of chimpanzee, *Pan troglodytes schweinfurthii* (the Belgian chimp), and this strain had already been designated as an outlier of the HIV-1 viral group.[182] Again, three out of four of these viral strains derived from chimpanzees consisted of only single genomes reconstructed from viral fragments. Based on this information, the investigators stated: "We conclude from our results that *P. t. troglodytes* is the natural host and reservoir for HIV-1."[136] This conclusion claimed instant international acclaim. On the same day the scientific article was published (February 1, 1999), this story was front page news in the *New York Times* and was also featured in a British Broadcasting Corporation (BBC) broadcast.[190, 191]

Interestingly, these investigators also listed reasons that "cast doubt on chimpanzees as a natural host and reservoir for HIV-1"; namely, "a wide spectrum of diversity between HIV-1 and SIV_{CPZ}, an apparent low prevalence of SIV_{CPZ} infection in wild-living animals, and the presence of chimpanzees in geographic regions of Africa where AIDS was not initially recognized."[136]

It is unfortunate that the viral strains from Marilyn and the Gabon chimps were phylogenically closest to HIV, since the natural range of *Pan troglodytes troglodytes* has no geographical proximity to the regions of Africa believed to possess AIDS endemicity. The natural range of the other chimp, *Pan troglodytes schweinfurthii*, does overlap these African areas of "AIDS endemicity," but the virus it hosts (derived from genomic fragments) qualifies only as an outlier of the HIV-1 group.[182]

"Remarkably, *these three HIV-1 groups were not each other's closest relatives* but instead were interspersed with SIV$_{CPZ}$ within the HIV-1/SIV$_{CPZ}$ radiation."[192] * (To the Author, the implication of this statement by this group of investigators is that the three major HIV-1 groups are, in fact, not closely related to one another at all. Just as all the various retroviruses isolated from African primates are related to HIV in that they are all retroviruses. By way of comparison, humans are related to gorillas, chimpanzees, and monkeys in that they are all primates. Humans also share skeletal homology with whales and dolphins. Yes how close are all these phylogenic relationships? The difference between humans, the lower primates, and aquatic mammals are obvious to any layperson; yet, the 'close relationship' of HIV to all these primate viruses are determined by virologists with vested interested in the outcomes of these comparisons.

In summary, it's conspicuously odd that another critical AIDS finding supporting the African AIDS origin theory involves a primate living in the United States. First, this chimpanzee theory is based on a single chimpanzee (Marilyn) that resided in the United States. Second, the incident of laboratory contamination that spawned the theory that HIV originated in African primates was due to a viral laboratory contaminant that was derived from sick Rhesus macaque monkeys that were also residents of the United States (as described in Chapter 5).

Nevertheless, this finding involving Marilyn begs to be reproduced; that is, as with the case of the mistaken identity, this single finding in a chimpanzee needs to be reproduced in separate laboratories by other scientists to be confirmed. Sadly, to date, this single finding is repetitively cited as the *de facto* substantiation for the origin of AIDS in *Pan troglodytes troglodytes*.

---

* *italics* added

Thus, the chimpanzee concept continues to receive substantiation and acclaim. In 2006, a team of researchers found SIV_{CPZ} antibodies and/or nucleic acids in roughly one-third of fecal samples from wild-living *Pan troglodytes troglodytes* in southern Cameroon. Using these data, the teams isolated and established the phylogenic relationship of 16 new viruses (amplification of *pol* and **env** genes); thereby, claiming to establish the progenitors of HIV-1 Groups Major and New within distinct wild chimpanzee populations.[193] For some reason, there is a strong psychological imperative to explain the origin of AIDS, so these findings were echoed by BBC on May 25, 2006,[194] and, no doubt, by others as well.

In another study, the *env* sequences of 3 wild-born Cameroonian chimpanzees were found closely related to those of HIV-1 Group N, discovered in the same geographic region, adding to an extremely small subset of viruses categorized as Group N. None of these three Cameroonian sequences was clearly related to HIV-1 Groups M or O.[195]

One interesting finding is that HIV and three of the SIV_{CPZ} are **isogenic**, meaning they have the same genomic backbone. HIV-1 and these three SIV_{CPZ} viral strains ($SIV_{CPZ-GAB-1}$; $SIV_{CPZ-GAB-2}$; and $SIV_{CPZ-ANT}$) all have one gene in common that is absent from all other SIVs, namely the **vpu** gene. Again, $SIV_{CPZ-ANT}$ remains divergent from all the rest.[192]

Again, the vast majority of primate immunodeficiency viruses cause neither illness nor immunodeficiency. Finally, in 2009 (24 years after Marilyn's death), a pathogenic strain of SIV was isolated in wild Tanzanian chimpanzees. At that time, 40 strains of SIV had been isolated among a number divergent captive and wild species.[154] Sadly, investigators persist in researching retroviruses in animal species far removed from the epicenter of the HIV/AIDS epidemic, rather than investigating the multitude of animal species in the United States that harbor retroviruses.

PART III

The African Fallacy

PART III

The African Fallacy

In essence, the purported African AIDS epidemic is described as a holocaust — a heart-rending loss of life on a massive scale. The African AIDS epidemic-as-holocaust is a mental construct based on (1) the inappropriate usage and inappropriate interpretation of first-generation tests for the HIV antibody, (2) a belief in the African green monkey (and/or chimpanzee) theory of HIV origin, (3) the use of a provisional surveillance definition for AIDS that lacks all laboratory tests, (4) a belief in the promiscuity of heterosexual Africans, and (5) the unspoken myth that there are no homosexuals in Africa.

The momentum of this holocaust belief is such that greatly exaggerated statistical projections persist today ("tens of millions infected . . . entire countries doomed . . ."). While statistical hyperbole has abounded throughout the history of the AIDS epidemic, sober reports of actual surveillance findings have been far less prevalent (and never sufficiently amplified). Surveillance reports tabulate the actual numbers of AIDS cases reported to WHO by each nation's health agency. Actual surveillance reports depicted a far different scenario than that of the widely reported, and erroneous, statistical analyses. For example, the November 26, 1999 issue of *Weekly Epidemiological Record* (a publication of the World Health Organization) reported that Africa had 794,000 AIDS cases. The United States had 717,000 AIDS cases in the same report. To place this data in the proper context; that is, *per capita*:[196, 197] *

- Africa   103 AIDS cases per 100,000

- United States   259 AIDS cases per 100,000

Thus, at that time, HIV prevalence in the United States exceeded that of Africa by a factor of 2.5. Moreover, these African surveillance data likely were inflated because of the use of the provisional surveillance definition (described Chapter 13) to define African AIDS

---

* values are rough estimates calculated by Author using 2003 U.N. population estimates

cases.[198, 199]

Occasionally, shards of clarity emerge from the maelstrom of hyperbolic reports. For example, on July 6, 2007, National Public Radio reported that almost 30 countries had reduced AIDS prevalence estimates . . . part of a worldwide trend. India has almost halved its internal AIDS estimate. (Moreover, African countries were essentially accused of inflating AIDS figures in order to receive additional funding.)[200]

In 2007, the Joint United Nations Programme on HIV and AIDS (UNAIDS) also reduced its estimate of people worldwide living with AIDS down to 33.2 million, from 39.5 million the previous year. The revisions were due mainly to improved methodology, better surveillance by countries, and changes in the key epidemiological assumptions used to calculate the estimates. Approximately 70% of the difference was explained by the reduction of prevalence in India and several sub-Saharan African countries, including Nigeria, Mozambique, Zimbabwe, Kenya, and Angola.[201]

Meanwhile, the population of Africa continues to climb. According to UN statistics, the population of Africa grew from 479 million in 1990 to 965 million in 2007.[202] No doubt, AIDS exists in Africa and everywhere else in the world to some extent, but the African AIDS epidemic-as-holocaust never manifested.

Chapter 11

The First African Patients – Diagnosed in Belgium

The first cluster of African AIDS patients was reported in 1983 – from Belgium.[203] This first cluster contained two males and three females. Two of them had died. All had denied homosexuality or drug use. They presented with prodromes of fever, weight loss, and generalized lymphadenopathy. Extensive investigations revealed no evidence of cancer (cancer and/or its treatment could possibly induce immunosuppression).[203]

None of these African patients had *Pneumocystis carinii* pneumonia (PCP) or Kaposi's sarcoma (KS). In the United States, these were two of the most common, early onset, AIDS-related, opportunistic infections. By comparison, among the first 108 AIDS patients in the United States, all had either PCP, KS, or both PCP and KS concurrently, and only one patient was female.

Two of the African patients had bacterial septicemia (*Salmonella typhimurium*), bacterial infections found in AIDS patients but not particularly indicative of early AIDS onset. Other fungal infections were present in these two patients, opportunistic fungal infections being more typically representative of HIV-infection. Two patients had herpes virus infections. One had **CNS** (brain) toxoplasmosis, a disease common in AIDS patients. All had poor **CD4/CD8 ratios**.

All patients were of "good" socioeconomic status. Three of them had been residents of Belgium for a range of 8 months to 3 years. Per the authors of this report: "This preliminary report suggests that Black Africans, immigrants or not, may be another group predisposed to AIDS."[203] At the time these first 5 African patients were diagnosed in Belgium, the United States had over 1200 reported AIDS cases.[47]

By 1984, this Belgian cluster had expanded from five to eighteen patients. The thirteen newcomers had more classical early onset AIDS-related diseases and classical AIDS-related immunological profiles. One-third of the newcomers were females. All of these patients were residents of Belgium, Rwanda, or Zaire. (Rwanda and Zaire were formerly Belgian colonies.) The African residents of Belgium traveled regularly to their country of origin.[59] This cohort would grow to 40 patients, one patient among them reportedly first seen in May 1979, concurrent with the foremost AIDS cases in the United States. Of these forty patients, only two had self-reported high-risk factors (two gay men).[204]

The physicians who first recognized and treated these patients wrote: "It is possible that AIDS has always been present but unrecognized in Africa. However, we are struck by the increasing number of patients who have come from Zaire or Rwanda to Belgium during the past 4 years to seek medical care. We believe that AIDS is a new disease that is spreading in Africa."[59]

"Homosexuality is a taboo subject in equatorial Africa, as it is in Haiti, and it is hard to get information about an African's sexual life. We are treating a Belgian homosexual who has AIDS. He has been living in Zaire for more than 20 years and has had many African sexual partners, nearly all of them bisexuals who engaged in homosexual relationships for *material profit* [italics added]. He also had sexual partners in Europe and in Brazil, and had numerous European friends in Zaire who had had similar homosexual relationships with black Africans."[205]

From *their* point of view, a lot of AIDS patients were coming from Africa seeking medical treatment in Belgium: so these physicians conjectured whether these patients might have acquired AIDS in their country of origin.[59] At the time these eighteen African AIDS patients (one-third female) were reported, the United States had 3000 AIDS patients (only seven percent female). Naturally, New York City was prominent in the number of AIDS cases (forty-two percent of all reported U.S. AIDS cases at that time).[14]

Oddly, shortly thereafter, when Europe had approximately 223 AIDS cases, twenty-six percent of these AIDS cases were reported among Africans.[206]

From Belgium to Zaire

The African residents of Belgium had traveled regularly to their country of origin. Most of them were from Zaire, a former Belgian colony, and some patients diagnosed in Belgium had come from Zaire seeking medical treatment.[59] These Belgian discoveries prompted a 3-week investigation in Kinshasa, Zaire that identified 38 presumed AIDS patients having a male-to-female ratio was of 1-to-1.1, with a slight predominance of females. All patients had been newly admitted hospital patients in Kinshasa, Zaire.[*] All the Africans denied homosexuality, IV drug use, or any history of blood transfusion. Although all socio-economic classes were represented, a disproportionate number of cases seemingly occurred in the higher income demographic.[58]

Overall, the clinical profile of these hospitalized Zairian patients differed from that of European and American AIDS patients. The most striking features were profound weight loss (> 10% body weight) and severe chronic diarrhea. All patients exhibited weight loss and 84% had diarrhea. Fever was also prominent. Oral thrush was present in 31% of the population, a prevalence similar to that of the US population. Something new and never seen before with AIDS: four women had **amenorrhea** (the cessation of menstruation). The development of amenorrhea might be explained by profound weight loss. It's a condition found among professional ballerinas and female anorexics as a consequence of their extraordinarily low proportion of body fat. Twenty percent of this cohort also developed **pruritic papular** skin lesions (hive-like rash). The possibility of cancer in these Zairians could only be excluded by clinical conditions: no laboratory evaluations were possible. Overall, this presentation profile differed from that of classical AIDS, meaning the presentation of AIDS as observed in the United States and Europe.

Conditions reminiscent of **classical** AIDS were also present. Half the population of

---

[*] *Republic of Zaire* is the former name of the *Democratic Republic of the Congo*

Zairian patients had generalized lymphadenopathy, which seems relatively low given that lymphadenopathy had been a key feature of U.S. AIDS patients (and usually a component of the AIDS prodrome).[207] Nine patients had anogenital *herpes simplex* infections, five patients had disseminated KS, and all patients had severely low CD4/CD8 ratios.[58] Ten patients died during the 3-week observation period.

Separately, outside of the hospitals, the investigators discovered two other small heterosexual clusters linking Belgium and Zaire: one cluster linked one male to four females, and the second cluster linked one female to three males. Most members of these two small clusters had been diagnosed in Belgium.[58] At this time, the apparent evidence of heterosexual AIDS transmission with a slight female predominance in population seemed particularly important.

From Belgium to Rwanda

Rwanda is another former Belgian colony. A similar 4-week evaluation was undertaken in Rwanda. Questionnaires were sent to all clinicians at the central hospital in the Kigali, the capital city, and a number of patients were evaluated. In this 4-week period, seventeen potential AIDS patients were identified, including two children. Nine other patients were reported as having AIDS prodrome ($n = 26$).[57]

Overall, the clinical spectrum was not particularly reminiscent of AIDS as observed in U.S. and European cohorts. All patients presented with generalized lymphadenopathy and severe weight loss (12–32% of body weight), and two-third had chronic diarrhea. One man had disseminated KS, two women had cutaneous KS, and about seven people had a combination of *herpes simplex* and/or problematic infections induced by organisms including tuberculosis, a fungus-like bacterium. Tuberculosis can be contracted without having HIV infection, but its clearance requires a competent cell-mediated immunity, which HIV disrupts; making tuberculosis resistant to cure. Candidiasis was present in almost half the group.

Salmonella infections were present also. Since these infections are bacterial, they can be present in AIDS patients but are not an early indication of HIV infection, particularly. The adult patients all denied homosexuality and drug use; however, thirteen out of seventeen men reported frequent contact with prostitutes, and three of these seven women were prostitutes.

As with the previous study and the African AIDS population in Belgium, almost all patients belonged to the middle or upper class socio-economic strata. Twenty-one of the 26 patients lived in urban centers.[57]

ORAL CANDIDIASIS

In a group of 37 seronegative patients with Kaposi's sarcoma, 30 of them had oral candidiasis. These patients obviously had some form of impaired cell-mediated immunity, but apparently not HIV infection.[208]

The Literary Trilogy

Together, these three reports from Belgium, Zaire, and Rwanda comprise the aforementioned Trilogy – the three pivotal publications forming the theoretical foundation for the presence of endemic AIDS in Africa.[57-59]

Consistently, throughout the next three or four years, virtually *every* medical or scientific report would cite these three authors as substantiation that AIDS was endemic in Africa. Over time, these citations would assume the connotation that AIDS originated in Africa.

By July 1984, European nations had reported 421 AIDS cases to the World Health Organization. Europe was defined as Denmark, France, the Federal Republic of Germany, Greece, Italy, the Netherlands, Spain, Sweden, Switzerland, and the United Kingdom. Among these 421 European AIDS cases, 17% of them were foreigners, namely: [209]

- 39 patients from Africa

- 17 patients from Haiti

- 16 patients from countries ranging from the Dominican Republic to Pakistan

Concurrently, the United States had 2200 AIDS cases.[18]

These manifestations of AIDS and/or other medical conditions were diagnosed before there was widespread capability to test for antibodies to HIV. When HIV antibody testing was implemented – first by laboratory-based investigations and then by commercial HIV assays – the next layer of error was added upon the previous misconceptions of an "epidemic" of African AIDS.

ROCK & ROLL JOURNALISM

A South African journalist, Rian Malan, wrote a revealing article about AIDS in South Africa, published in *Rolling Stone*, November 2001. After a year of journalistic research, and discussion with experts of divergent viewpoints, he could only consider himself stymied as to whether "one in four" South Africans could currently be infected with HIV, as was widely reported.

Yet Malan reported at least two interesting findings. First, he attempted to confirm a newspaper account that a new paper-coffin maker was overwhelmed with orders. Instead, he found the paper-coffin maker and his competitor had gone out of business. The national industrial coffin makers reported: "It's quiet." "We aren't feeling anything at all." "If you read what is in the papers, we should be overwhelmed, but there's nothing. So what is going on? You tell me."

Malan also called WHO in Geneva to confirm a set of statistics that he found difficult to reconcile. According to WHO estimates, 400,000 South Africans died from AIDS during 2000. However, according to the South African Department of Home Affairs, there were only 457,000 registered deaths for the entire country in 2000. Taken literally, these figures mean 85% of deaths in 2002 were attributable to AIDS whereas only 25% of the population was purportedly infected.[210] All this is highly questionable.

Chapter 12

HIV Antibody Testing – An Avalanche of False-Positives

All the aforementioned AIDS cases were diagnosed prior to the availability of commercial HIV antibody tests. The first commercial HIV antibody assay (**ELISA**) was licensed in 1985, making HIV antibody testing available to doctors in clinical and hospital settings and blood donation facilities.

Prior to the availability of the commercial assay, methods of detecting HIV antibodies were complex laboratory procedures best suited to scientific research settings.[211, 212] Nevertheless, a few intrepid investigators attempted to utilize these laboratory methodologies for the screening of both symptomatic and asymptomatic (apparently healthy) populations.

The Avalanche

Zaire became the center of attention . . . and the results were astounding! In Zaire, twelve to twenty-four percent of the "rural poor" tested positive for the HIV antibody ($n = 250$). Although none of these rural poor Zairians had any symptoms suggestive of AIDS or HIV infection, twelve percent were "clearly positive" while another twelve percent had borderline ratios.[213, 214] Comparatively, at this time, about 10,000 AIDS cases had occurred in the United States (approximately 0.00004% of the U.S. population).[44] The authors of the study concluded that either HIV or a *closely related, cross-reactive virus* might be endemic in the region.[213] * Of course, such *qualitative phrases* tend to get lost in translation.

Such vignettes were replayed in numerous African countries; the findings not necessarily reported as front page news, yet were repetitively cited throughout the medical literature, becoming the common language of reportage.[215, 216]

Astutely stated by one team: "Preliminary serological studies based on [ELISA] suggested a wide and early presence of HIV in many regions of central Africa, sometimes even

---

in regions where AIDS cases have so far not been observed."[217] Thus, high percentages of

asymptomatic populations tested seropositive for HIV, including: [218-221]

Seroprevalence	Description	No.
12.5 %	Healthy Zairians, Residents in Belgium	40
15.4 %	Healthy Adults in Kampala, Uganda	716
15.5 %	Rwandan Blood Donors	51
20 %	Healthy Ugandan Residents	51
27 %	Zairian Prostitutes	377
80 %	Rwandan Prostitutes	84

In another striking report, sixty-six percent of stored blood samples taken from

Ugandan children in 1972 and 1973 were borderline seropositive.[222] The test results of these

children's blood and other stored blood samples supported the belief that HIV had been

endemic in Africa for years, "existing in a population acclimated to its presence," [222] the

biological analogy being that AIDS was like mononucleosis: a viral pathogen whose exposure

during childhood yields an asymptomatic infection whereas delayed exposure to naïve adults

results in serious morbidity. The idea was that HIV had been endemic among rural populations

in Africa for 30 or 40 years before ecological and sociological changes forced it out of the

jungle into to naïve, highly susceptible, urban populations wherein the disease spread

exponentially.

Less than a year later, report came of HIV antibody testing in an urban population of

medical patients in Kinshasa, Zaire, a population purported to have AIDS ($n = 332$). Ninety-

nine percent tested HIV seropositive by a commercial ELISA and Western blot confirmatory

test. They were also marked by inverted CD4/CD8 ratios. As with the aforementioned

investigation of 38 purported AIDS patients in Kinshasa, the male-female case ratio was 1:1.1

with a slight predominance of females.[223]

Although current HIV testing kits sometimes still generate false-positive and false-

negative results, as with all other diagnostic or screening assays commonly applied in clinical

or hospital settings. Later generations of these HIV antibody assays would improve in accuracy. Therefore, in 1994 (ten years after the first tests of the rural poor), a researcher tested the blood samples of 250 Zairian patients that had been stored since 1969. This time, the stored blood samples were evaluated by three separate techniques representing 1994 state-of-the-art assay technology: (1) a current generation ELISA, (2) the confirmatory Western Blot test, and (3) polymerase chain reaction (PCR) — a stringent and perhaps unprecedented evaluation. As reported by the researcher: "Interestingly, none of the patients was confirmed positive for HIV, even though this region is now endemic for HIV-1."[224]

ENZYME-LINKED IMMUNOASSAY (ELISA)

The ELISA test is used to screen asymptomatic blood donors and diagnostically evaluate patients who have been pre-screened by risk factor assessment and/or clinical presentations. Currently, assay architectures are configured for a variety of applications.[225] In clinical settings, meaning hospital laboratories and laboratory testing services, the assay process is automated and therefore objective; patient blood vials are simply inserted into the machine.

ELISA test results are not "black and white"; rather, test outcomes are a gradation of gray, or more accurately, a gradation of color change. Some specific level of color density is designated as "positive." This measurement is performed automatically by a spectrophotometer. This positive value is reached through trial and error and this positive value varies from population to population.

The ELISA test is very sensitive, meaning it detects small amounts of antibodies. Being so sensitive, ELISA is also subject to being fooled by the presence of proteins or antibodies unrelated to HIV. A high percentage of ELISA seropositive results are false-positives. Another procedure was required to confirm that these antibodies in question belong to HIV.

Not all authors were so baffled, as expressed in the words of one unsung author: "The first generation of serological tests for anti-HIV-1 gave so many false positives with African sera that it was wrongly postulated that the virus was endemic in Africa. As there is no simian or other virus sufficiently closely related to HIV-1 as to suggest a recent common ancestor, the evolution of HIV-1 is obscure and there is no current evidence to support the hypothesis of an African origin."[226] Such opinions were never echoed in the general media.

The Lack of Avalanche

The avalanche of false-positives was a compelling story whose amplification through repetition became firmly established in the psyche of the medical community and laypeople alike. As always, the subsequent corrections and antipodal findings received far less attention. As one investigator stated: ". . . [subsequent] investigations based on more specific tests such as radioimmunoprecipitation, competitive radioimmunoassays, or competition ELISAs could not confirm some of these early tests."[217]

For example, in October 1985, an investigative team reported testing the serum samples of 2128 Africans for the HIV antibody — only 2 patients were seropositive, and they were the only patients in the series with clinically diagnosed AIDS.[227] (Interestingly, 12 other "AIDS suspects" tested negative.)[227]

In this study, two rounds of antibody testing were performed. First, all samples were tested with a laboratory-based, enzyme-linked immunoassay (ELISA). Second, the samples were tested with a technique similar to the Western blot (SDS-PAGE).* Although only 2 of 2128 samples would eventually test positive, there were 348 false-positives (approximately 16%) during the first round of testing with the ELISA method, as follows: [227]

Description	ELISA[v]	SDS-PAGE
Healthy Africans (*n* = 1794)	287	0
Sick Africans (*n* = 344)	61	2

[v] under these circumstances, a positive ELISA outcome equals a false-positive

This population consisted of 1794 healthy Africans from Senegal, Nigeria, Liberia, Gabon,

---

* SDS-PAGE – polyacrylamide gel electrophoresis

Zaire, Kenya, and South Africa (serum samples collected in 1981); 334 clinic attendees in Lambaréné, Gabon; 63 leukemia/lymphoma patients, and 14 suspected AIDS patients. It must be granted that among these locales, only Kenya and South Africa were purported to have massive epidemics at various times, except for the West African countries wherein HIV-2 has emerged (HIV-2 being a matter to be discussed in its own right, in a separate report). It is interesting to note that only 2 of 14 AIDS suspects tested HIV seropositive.

Per the investigators: "Our low frequency of HTLV-I and HTLV-VIII positivity, with the specific immunoprecipitation method applied to sera showing ELISA positivity, suggests that African sera yield a high prevalence of false-positive reactions with the ELISA for HTLV-I and HTLV-III."[227] In addition, these authors evidently believed that the failings of the ELISA alone accounted for the extraordinary amount of false-positives being reported by other researchers; discounting the possible existence of the theoretical, cross-reacting, endemic retrovirus speculated by other authors.[227]

A year later, a research group reported testing the serum samples from 6015 healthy Africans collected between 1978 and 1984 — only 4 samples tested HIV seropositive. The two tables below represent the two different sub-groups in these six thousand Africans. The first test group was evaluated by both ELISA and immunoprecipitation (IP) ($n = 2573$). In this population, only 2 serum samples from healthy Africans plus 2 serum samples from AIDS patients tested seropositive, as shown in Table 14.[228] It is interesting to note that 130 false-positive ELISAs occurred among 552 hospital attendees, suggesting that people with morbid conditions have high false-positive ELISA rates, although other factors might be at play.

Table 14 – Africans tested for HIV antibody by IP and ELISA (*n* = 2573)

Subjects	Years of Sampling	Country	No. of Samples	ELISA Positive	IP Positive
General Population	1981	Senegal	789	66	1
Hospital Attendees	1982	Nigeria	236	9	0
Blood Donors	1984	Nigeria	300	0	0
Pregnant Women	1983	Gabon	144	1	0
Hospital Attendees	1983	Gabon	552	130	1
Villagers	1983	Gabon	98	8	0
Hospital Attendees	1983	Zaire	15	1	0
Students	1981	Kenya	300	16	0
Hospital Attendees	1984	Kenya	133	7	0
AIDS suspects	1985	Guinea-Bissau	6	2	2
TOTAL			**2573**	**240**	**4**

The second test group was evaluated by immunofluorescence (IF) assay (*n* = 3464). In this population, only 2 samples from the general population tested positive. Thirteen AIDS patients and two of their sexual partners also tested seropositive, shown in Table 15. [228]

Table 15 – Africans tested for HIV antibody by IF and ELISA (*n* = 3464)

Subjects	Dates of Collection	Country	No. of Samples	No. Positive
Villagers	1976-84	Liberia	935	0
Villagers	1978	Ivory Coast	1195	0
Villagers	1983	Burkina Fao	299	0
Clinic attendees	1981-4	Gabon	855	2*
Villagers	1983	Uganda	164	0
AIDS Patients	1985	Zaire	13	13
Sexual Partners	1985	Zaire	3	2
TOTAL			**3464**	**17**

* one sample was from a white subject

Another investigative team subjected 286 African serum samples collected in rural Tanzania between 1982 and 1984 to a panel of antibody tests: an ELISA performed at the U.S. National Institutes of Health and two commercial ELISAs. Eighteen percent (18%) of the samples tested positive in the NIH ELISA; yet all samples tested negative by the two commercial ELISAs.[217] * In addition, these researchers concluded: "The Western blot reactivity pattern of these individuals was completely different from that of European or American HIV-infected individuals." The Tanzanian sera had a much lower affinity for HIV **gag proteins** (p17, p24, and p55), and a disequilibrium or absence of specific HIV antigens compared to that of classical AIDS patients. The researchers concluded that the most likely explanation for the initial ELISA false-positives was an *unknown retrovirus or cross-reactive agent* (italics added).[217]

Some other small studies had similar findings:

- No sera samples from Cameroon tested convincingly positive ($n = 375$), although some people might have been exposed to a virus "*related to but not identical with [HIV]* (italics added)." [229]

- No old people in Uganda tested positive ($n = 96$); suggesting that HIV did not exist in the region historically.[220]

- No pygmies tested positive ($n = 340$). Pygmies reportedly come into close contact with monkeys, whom they hunt for food; live a forestial existence in the North Congo and Central Africa Republic (i.e., Central Africa, supposedly the center of AIDS endemicity); and reportedly had no sexual congress with neighboring populations.[230]

- No sera samples from the mangrove and swamp zone of eastern Nigeria tested positive ($n = 301$) as evaluated by laboratory-based ELISA and immunoprecipitation (IP). [231]

- None of 677 sera collected from children between 1964 and 1975 in Uganda, Central Africa, and North Africa were seropositive for HIV, per laboratory based immunoblot testing (similar to Western blot).[232]

---

* 52 samples were positive per NIH ELISA and 49 of these were re-tested by commercial ELISAs

HIV/AIDS – Facts & Fiction   Page 81   Copyright © 2012 by Chris Jennings

- Of 2574 sera collected in Mozambique, Malawi, South Africa, Lesotho, Botswana, Angola, and Swaziland between 1970 and 1974, none were found positive by Western blot; however, one commercial assay detected 11 false-positives, another commercial assay detected 6 false-positives, a third commercial assay reported zero positives, and a fourth commercial assay yielded an unquantified "many false-positives."[233]

- Of 640 Nigerians evaluated by ELISA and immunoprecipitation (IP), none were seropositive by IP, although 12 (1.9%) were seropositive by ELISA.[234]

Most of these findings were reported in the medical literature in 1985 and 1986, basically concurrent with the hyperbolic reports of an avalanche. But none were amplified by the general and medical media; therefore, none reached the general consciousness of the medical and lay public.

The Pitfalls of Testing

Technically, the early generation laboratory and commercial assays used in these initial surveys lacked "specificity," meaning they cross-reacted with extraneous antibodies and other biological entities, yielding an exceptional number of false-positives.

Any number of factors is able to generate false-positives in HIV antibody tests. For example, medical conditions other than viral infection can induce false-positives. Such conditions included hematologic malignancies, DNA viral infections, autoimmune disorders, myeloma, biliary cirrhosis, alcoholic hepatitis, renal transplantation, hemodialysis, connective tissue disease, and second or subsequent pregnancies.[235-245]

Moreover, with the laboratory-based assays, analytical outcomes would vary by laboratory. Each investigative team assembled their own ELISA platforms by following published procedures, sometimes borrowing HIV isolates and/or cell-lines.[61, 211, 228, 246, 247]

(Just as a pie tastes different when cooked by different people though using the same recipe and ingredients.) In the same vein, commercial products would vary by manufacturer in regards to their viral antigenic contents, HIV propagation cell-lines, and consequent antigen sensitivity.[217, 248] Lots within manufacturers might also vary, and identical testing kits yield different rates of false-positives at different laboratories.[249, 250]

In addition, there were episodes wherein **immunoglobulin** products passively transmitted the HIV antibody to patients (the immunoglobulin contained the HIV antibody only, not the virus itself; the antibodies faded away over time since no antigens were present).[251-255] Some vaccine products also induced false-positive reactions in the HIV ELISA and other assays: one to two percent of the vaccine recipients had non-specific antibodies that reacted with the polystyrene (plastic) components of the test apparatus.[242] Yet given all these factors together, it remains difficult to explain the breadth of seroprevalence reported among apparently healthy Africans. Other confounding factors were likely at play.

Regarding HIV assays in particular, the market needs are dominated by the blood testing services. As stated by WHO: "To serve the needs of blood transfusion services, which use the vast majority of all HIV tests worldwide, increasingly sensitive HIV antibody assays have been developed in order to shorten the window period (the interval between the point of infection and the development of detectable antibody). As a result of this trend, less sensitive but highly specific HIV tests have been withdrawn from the market."[256] In this setting, the words "increasingly sensitive" means the assays generate more false-positives under these circumstances.

Western Blot – The Confirmatory Test

In standard clinical settings, repeatedly reactive ELISA samples undergo confirmatory evaluation by Western blot. Unlike the entirely automated ELISA, the portions of the Western blot are performed manually, and the test results must be interpreted by a technician. Whereas the ELISA is a sensitive assay, the Western blot is a very *specific* test: it is theoretically very exact in what it "sees," thereby correcting most of ELISA's mistakes. Western blot kits are also commercial products.

In the Western blot, blood samples are exposed to a gelatin strip in which HIV viral proteins have been embedded. If HIV antibodies are present in the blood, then they bind to the HIV antigens embedded in the gelatin strip; creating a characteristic banding pattern. Depending on the stage of HIV infection, one or more bands may be visible. The results must be interpreted by the technician, and are based on the appearance of the banding pattern, current standards of diagnostic interpretation, and experience. As such, Western blot test results are subjective judgments.

HIV Antibody Tests at the American Red Cross

By way of comparison with the seroprevalence reported in tropical settings, the American Red Cross reported the following false-positive rates among 2.58 million blood donors in 1986: [257]

Number of Blood Donors	2.58	million
ELISA positive	1	%
ELISA positive repeatedly	0.27	%
Western blot positive	0.035	%

Antigens

Initially, both laboratory and the first-generation commercial tests were **whole virus assays**.[211, 212, 248, 258, 259] Whole viruses were cultivated, extracted, and disrupted ("crushed"). All disrupted parts of the virus – proteins, enzymes, genetic material – were contained in the end-product of this process, the **viral lysate**. In this lysate, the surfaces of the viral protein components were covered with the viral antigenic determinants. The viral lysate is utilized as the reactive surface of the assay, the surface that captures HIV antibodies. A basic tabulation of HIV's viral components is presented in Table 16.[248, 260, 261]

Table 16 – Select HIV Genes & Antigenic Gene Products

Gene	Gene Products	Viral Component	Description
ENV	gp160 gp120 gp41	Envelope Proteins	Env precursor (gp120/p41) Envelope protein Transmembrane protein
GAG	p55 p24 p17	Core Proteins	Gag precursor Capsid protein Matrix protein
POL	p64 p53 p31	Enzymes	Reverse transcriptase Reverse transcriptase Integrase/Endonuclease

At first, a core protein (p24) and the transmembrane protein (gp41) were recognized as the most reactive HIV antigens, forming dark bands on the gelatin strip.[212, 257] Before 1987, by CDC and NIH standards, the finding of p24 or p41 alone was considered a positive confirmation of HIV infection.[248, 262-264] As late as 1989, a finding of either p24 alone and/or in conjunction with p41 was considered a positive confirmation per the recommendations of some manufacturers and expert authorities.[235, 262, 265]

Also, some early laboratory assays contained purported HIV proteins of the molecular weights that are not standard by today's measure.[223, 266, 267] (One must allow some imprecision due to variation in laboratory procedures and measures, but other non-HIV viral particles might have been present.[*])

In later generations of both ELISA and Western blot, the envelope precursor glycoprotein (gp160), the envelope glycoprotein (gp120), and the transmembrane protein (gp41) were required for a confirmatory positive finding, usually in conjunction with one or more other antigens.[261, 268] But such practices were not reliably encouraged until 1996.[269] (The precursor glycoprotein, gp160, is cleaved by a protein to create gp160 and p41. Glyco- means "sugar.") Currently, assay architectures are configured for a variety of applications.[225, 270]

The HIV antigens used for these assays were not pure. They also contained **cellular antigens** from the cell lines in which HIV was cultivated.[**] Generally, the human cell line H9 was used. Therefore, the viral lysate also contained human cellular antigens. The blood of some people contains anti-human antibodies that induce false-positives in the ELISA.[236, 257, 269, 271-275] (Other non-viral, cell-derived proteins, such as ribonucleoproteins, may also cause problems in the same manner.[257]) People who have been repetitively exposed to other people's antigens frequently have anti-human antibodies. Such people include blood transfusion recipients, organ transplant recipients, IV drug users, people infected with tapeworms, and multiparous mothers (i.e., pregnant women having one or more prior pregnancies).[236, 248, 276-279] A problem with false-positivity has been noted among pregnant women in a variety of settings:

- In 1991, Russia had 30,000 false-positives out of 29.4 million tests: 8000 of them pregnant women. Only 6 women were confirmed positive.[280]

- Between 2004–2006, two Tokyo maternity hospitals evaluated 6461 pregnant Japanese women: 27 tested positive; 1 was confirmed positive.[281]

The overall lack of specificity of the ELSA in pregnant women was demonstrated in a small study in which the positive predictive value of the ELISA was measured in small populations of pregnant women. (The **positive predictive value** is the probability that a

---

[*] Author speculation
[**] "cellular antigens" here means Human Leukocyte Antigens (HLA), the molecular flags that designate "self"

patient with a positive test result really does have the condition for which the test was conducted.) The positive predictive value (**PPV**) for an ELISA was only 9.8% and 35.7% for Hispanic women (U.S. residents) in select populations. In these populations, Hispanics comprised the greatest percentage of the population but had the lowest the rate of disease. They were a low-risk population. The corresponding positive predictive values for the African-American and Caucasian patients in the same environments were much higher, as seen in the table below (a higher Disease Index indicates a higher rate of confirmed HIV infection): [282]

	Disease Index*	PPV	% of Population (n = 9,781)**
African-Americans	2.43	82.6%	8%
Caucasians	1.53	75%	2%
Hispanics	0.05	9.8%	87%
Asians	--	--	3%

*per confirmed HIV seropositivity; **9,781 deliveries, 69 initial positives, 26 confirmed

This data set embodies some well-recognized facts about medical assays. First, medical assays are most accurate in a diseased population. Second, a medical assay is likely to generate more false-positives than true-positives in a large healthy asymptomatic population.[235, 283] Consequently, very few medical assays are approved for screening large asymptomatic populations; rather, these assays are intended to confirm or substantiate a disease condition indicated by prior clinical and/or laboratory work-up.

Regarding HIV assays in particular, the market needs are dominated by the blood testing services. As stated by WHO: "To serve the needs of blood transfusion services, which use the vast majority of all HIV tests worldwide, increasingly sensitive HIV antibody assays have been developed in order to shorten the window period (the interval between the point of infection and the development of detectable antibody). As a result of this trend, less sensitive but highly specific HIV tests have been withdrawn from the market."[256] In this setting, the words "increasingly sensitive" means the assays generate more false-positives under these circumstances.

Nevertheless, the most widely available epidemiological data on HIV/AIDS in Africa are seroprevalence data.[284, 285]

In Africa, a large percentage of seroprevalence studies were conducted in prenatal clinics.[226, 286, 287] In these surveys (1987), WHO carried out serosurveillance surveys in Central African blood donors and pregnant women with seroprevalence rates ranging from 5% to 30%. Their Global AIDS Programme utilized these statistics in an AIDS projection model predicting a 30% reduction in the urban population growth rate.[288]

In recent years, the Republic of South Africa based its estimates of national HIV prevalence upon seropositivity findings in its antenatal clinics.[289] These annual surveys included only first-time pregnant women which should have precluded a higher false-positive rate found in multiparous women, yet HIV seroprevalence in South African first-time pregnant women reportedly increased from 22.4% in 1998 to 29.4% in 2009.[290-296]

Technical Issues

Cut-Off Level: In the ELISA, the antibodies are marked with a dye. The substrate, the testing surface of the assay, changes color as the antibodies accumulate. The idea is to capture antibodies specific to HIV antigens; however, there will always be some amount of binding by non-specific antibodies. The degree of color change is read automatically the spectrophotometer. The **cut-off level** – the positive value – is reached through trial and error, and this positive level varies from population to population. For example, from Boston to New York to San Francisco — each hospital setting is required to set its own numerical value for positive, established by substantiation of the assay outcomes through patient follow-up. This adjustment in positive values is a consequence of innate differences in these geographically (and perhaps demographically) distinct populations.

As stated by an industry standard: "We believe it is important to emphasize the importance of regional validation and verification of subclass partitioning as essential to the utility of a given laboratory's use of published reference intervals. The concept of a nationally or geographically universal reference interval, while perhaps desirable, is an externally difficult achievement."[297]

In the literature, investigators reported their cut-off values only intermittently. For laboratory ELISA methodology, the standard published guidelines followed by many researchers used an absorbance ratio of ≥ 2.1. In 1984–1985, the Author noted the use of a 5.0 cut-off ratio; one investigator having raised the cut-off level from 3.0 to 5.0. In 1986, components of a single report suggested that a cut-off ratio of 6.0 had become standard. Three years later, a laboratory reports using a cut-off of 1.0 wherein the laboratory standard was 0.3. These values may reflect changing patterns or the methodologies of different manufacturers and time periods.[211, 213, 214, 298-300] However, it is possible that one contributory factor in the initial high false-positive rate was the use of low cut-off values.[*]

Photospectrometer Filters: One lab reported that using two wavelengths (two filters) on their photospectrometer reduced the number of samples initially testing as HIV seropositive. Using one filter, they had forty-eight initial positives with only 8 repetitively positive. Using two filters, they had 10 initial positives with only 6 repetitively positive ($n = 1500$). [301]

Judging by Eye: In Venezuela, Western blot and radioimmunoprecipitation bands from malaria patients were similar to those of AIDS patients, including the major bands corresponding to p42 and gp120/l60.[302] [*] To paraphrase a team of Papuan New Guinea investigators: 'Given the malaria antibody generates strong immunoblot bands at p17 and p24 along with other strong bands not indicative of HIV, perhaps the intensity of color change could be misjudged as positive when inspected visually, rather than by expensive photometry, as might happen in poor-third world medical facilities.' [302] This is a speculation from afar, yet at least one other author found the Western blot banding patterns, of patients with malaria,

---

[*] Author's speculation
[*] all evaluations by indirect immunofluorescence, Western blot, and/ radioimmunoprecipitation

resembled those of AIDS patients.[302]

Freshly-Clotted Blood: Sera aspirated from freshly-clotted blood often gave rise to false-positive or equivocal results on a rapid commercial diagnostic assay. The situation was noted while screening organs for donor transplant in which knowledge of HIV status was required within 120 minutes. The investigator did not quantify the relative proportion of false-positives generated in this environment.[303]

Testing Bias: HIV testing results were strongly biased by the selection of specific African populations; specifically, the repetitive selection of prostitutes, patients in sexually transmitted disease clinics, and pre-natal clinics. All of these populations were based primarily in urban settings. Between 1985 and 1988, ten published HIV seroprevalence reports of African populations featured prostitutes as the principal or major component of the study population — without mentioning the word "prostitute" in the title.[57, 226, 304-311] An additional twelve HIV prevalence reports of African prostitutes had the word "prostitute" in the title.[61, 221, 312-321] This list is not necessarily exhaustive. Regardless, most of these prostitutes were healthy, overall (or lymphadenopathy being a common finding). Unfortunately, the seroprevalence findings derived from these select, biased populations tended to be extrapolated to the entire national and/or regional population(s) as a whole.

Concurrent Infections

In many ways, the most satisfying conclusion would be that some other non-HIV related infectious process confounded the HIV antibody tests, thereby giving rise to high percentages of false-positives. Naturally, given the circumstances, such an infection would be endemic to the tropics; something not found in temperate climatic zones.

Unknown Retrovirus: Several investigators in Africa and the unnoticed Amazon suggested that unknown endemic retroviruses might be confounding the HIV antibody tests.[213, 217, 227, 246, 258, 264, 322-324] Among retroviruses, the core proteins are highly conserved, meaning that different species of retroviruses might have viral core protein of similar molecular weights and antigenic properties as those of HIV. The enzymes found in retroviruses are also highly conserved, some of them being antigenic.[101, 164, 325] Antibodies to such theoretical viruses might fool the HIV antibody test. Retrovirus infections are considered rare in humans. For example, HIV was only the third retrovirus known to infect humans. The other two (HTLV-I and –II) were not very widespread in the United States and Europe; rather, they primarily existed in Africa and Japan. If such tropical retroviruses exist, then they must necessarily be benign and of little investigative interest: no investigative pursuits of such theoretical retroviruses have been undertaken.

THE AMAZONIAN AIDS EPIDEMIC

Per HIV antibody testing, in 1985, a secret and silent AIDS epidemic was raging in South America simultaneously with the rather publicized African version. Serosurveillance studies had shown that HIV antibody was endemic or epidemic in several Venezuelan populations, including several aboriginal, Amazonian, Indian tribes.

In one study, 3% of the Yanoama (n = 150) and 13% of the Ramon (n = 15) tested HIV seropositive.[▽] However, none of 211 randomly chosen, healthy blood donors from Venezuelan cities were seropositive.[323] In a separate serosurveillance, over 2% of a general healthy Venezuelan population tested HIV seropositive (n = 465).[‡] The HIV seroprevalence rate was 4% among patients with **Chagas disease**, and 29% among patients with acute malaria. None of over 169 randomly chosen, healthy blood donors from seven major Venezuelan cities were seropositive.[246] Both sets of investigators concluded that HIV or a closely related cross-reactive virus might have been endemic in this region.[246, 323] Conversely, a year later, an investigator found no confirmed positives in 556 samples from rural and urban areas in Venezuela (nineteen false-positives).[326] None of these investigations seemed to reach the consciousness of the medical and general pubic.

[▽] per indirect immunofluorescence test (IF); confirmatory by Western blot and radioimmunoprecipitation
[‡] per indirect immunofluorescence test (IF) and locally propagated cell line

---

Malaria: Malaria seems to have a questionable relationship with HIV seropositivity. One finding was repetitive but not always consistent . . . a high percentage of people with malaria tested positive for HIV antibodies: up to 42% of small malaria populations.[258, 299, 302, 327-330]* Conversely, groups of people infected with HIV tested falsely positive for the malaria antibody; such false-positivity rates ranging from 11% – 70% among gay men, hemophiliacs, IV drug users, and unspecified Caucasians.[328, 331]

Yet, paradoxically, within malaria cohorts, HIV seropositivity and malaria seropositivity had no statistical correlation; also, researchers couldn't find any cross-reactivity between *Plasmodium falciparum* and HIV. [258, 299, 327, 329]   Research in Africa, Venezuela, Bolivia, Thailand, Papua New Guinea, and the United States had failed to establish any valid statistical relationship.

One investigator suggested these false-positive outcomes were due to the non-specific activity of **anti-lymphocyte** antibodies, **anti-nuclear** antibodies (antibodies versus cell nucleus), and other **auto-antibodies** (antibodies versus self). These antibodies against self are common to patients with recurrent malaria. This investigator stipulated that this artifact had been removed from his assays; otherwise, they would generate an HIV seropositivity profile that "mimics that of endemic malaria."[332]

Several possible explanations were offered by another research team, the last being: "Finally, reactivity in both the ELISA and Western blot analysis may be non-specific in healthy Africans . . . we have emphasized in previous studies that the profile of reactivity in the Western blot analyses of ELISA-positive healthy subjects does not appear typical of seropositive American and European subjects and speculated that such reactivity may be related to other, as yet unknown, retroviruses in this environment."[258]

---

* patients with acute malaria and/or seropositive for antibodies to the malaria parasite (*Plasmodium falciparum*)

Biologically, it is well understood that the gradient of life increases as one travels from the global poles to the equator. Relatively speaking, fewer species survive near the north and south poles. The diversity and density of life – the number of species per square meter – increases as one approaches the equator. This fact is obviously true for animals and plants, and presumably so for micro-organisms, as well. In all likelihood, populations in equatorial tropical zones (such as Central Africa and the Amazon) are exposed to a far greater number and diversity of bacteria, fungi, and viruses than are human populations living in the temperate zones.

One concept, echoed briefly through the medical literature, seemingly is traced back to one comment made by an author in the New England Journal of Medicine. The comment was that Africans have "sticky" blood, meaning that the sera of Africans seem to induce a relatively high level of false positives.[333] No one seems to have quantified this effect, but perhaps this "stickiness," this high level of assay reactivity, results from the presumably higher level of exposure to uncharacterized viral infections or biological entities.

Malaria as Marker: Of course, it was thought that malaria might have been a **marker** for some other unknown biological entity that induced HIV false-positives. For example, malaria patients may receive transfusions more frequently than the norm, and blood transfusions carry the risk of HIV infection, as well as a risk for the development of anti-human antibodies.[334]

Among a cohort of children tested for the HIV antibody in Kinshasa, Zaire (1985), there was a "strong dose-response association between transfusions and HIV seropositivity. Of 167 hospitalized children, 112 (67%) had malaria, 78 (47%) had received transfusions during the current hospitalization, and 21 (13%) were HIV seropositive. In some settings, blood transfusions have been common treatment for children with malaria, who suffer from red blood cell concentrations. Ten of the 11 seropositive malaria patients had received transfusions during the current hospitalization; pretransfusion specimens were available for four of these children and were seronegative."[335]

It was naturally assumed that these children contracted HIV via blood transfusion. However an alternative explanation is that they merely converted to seropositivity because they had developed anti-human antibodies.

By one report, an "estimated 1 million transfusions [were occurring] per year in sub-Saharan Africa by 1970 and 2 million per year by the 1980s, indicating that transfusions were widely used throughout sub-Saharan Africa during the crucial period of 1950–1970, when all epidemic strains of HIV first emerged in this region."[336] Typically, these authors contribute seropositivity to HIV infection; whereas this widespread practice of blood transfusions could create substantial populations of people with anti-human antibodies.

Another theoretical possibility for these false-positives would be an unknown biological agent that travels the same pathways as malaria. Such a theoretical agent would presumably exist in the same environment as malaria and share transmission routes. Malaria is transmitted by mosquitoes which are **biological transmitters**, meaning that the mosquitoes themselves are infected by *Plasmodium falciparum* (the pathogen infects the salivary glands of specific species).

There is one known pathogenic retrovirus that is transmitted by insects: the equine infectious anemia virus (EIAV). Horseflies and deer-flies (but not mosquitoes) transmit EIAV between horses residing in close proximity. These insects are **mechanical transmitters**, meaning that their contaminated feeding parts transfer infected blood from one animal to another, but the infectious virus does not propagate within the insect itself.[337] One might say that equine infectious anemia virus is closely related to HIV. They both belong to the same genus, Lentivirus, but no evidence exists that HIV can be transmitted by mosquitoes or other blood-sucking insects. The equine infectious anemia virus is endemic in parts of the Americas,

Europe, Middle East, Russia, and South Africa.

Chapter 13

The WHO Provisional Surveillance Definition

Given all the reports of seropositive Africans permeating both the medical and general media, one could incorrectly suppose that American and European diagnostic algorithms were at play in Africa. (An **algorithm** is a step-by-step procedure for diagnosing medical conditions.) Quite the contrary, routine use of the ELISA HIV antibody tests for the diagnosis of AIDS remained restricted to the United States, Europe, and countries (or select medical venues) with similar medical economies.

In some regions of Africa, excluding elite urban hospital settings, the standard mechanism for diagnosing and reporting AIDS was the provisional surveillance definition. This diagnostic algorithm was designed by the World Health Organization (WHO) specifically for use in medical settings that lacked laboratory facilities for detecting HIV antibody or immunological dysfunction (CD4/CD8 ratios and such). The surveillance definition is used by national health authorities to systematically collect data on notifiable diseases.

As such, the *WHO provisional surveillance definition* relied solely on clinical symptoms; i.e., the symptoms seen by the doctor or reported by the patient. By the original 1985 definition, the *major symptoms* were: (1) weight loss of 10% or more; (2) chronic diarrhea for 1 month or more; and (3) prolonged fever for 1 month or more. The *minor symptoms* were: a cough persisting for 1 month, generalized pruritic dermatitis, recurrent herpes zoster, oral or pharyngeal candidiasis, herpes simplex infection, and lymphadenopathy.[198, 199] Any combination of two of these major symptoms plus any one of the minor symptoms equaled a diagnosis of AIDS. As such, weight loss of 10%, a fever persisting more than a month, plus a cough persisting for a month, or lymphadenopathy (swollen glands), equaled a diagnosis of AIDS. All of these symptoms can occur with or without HIV infection.

The WHO provisional definition, although praised in the medical literature, has also been criticized for likely misdiagnosing persons with pulmonary tuberculosis as having AIDS.[338-343] Such misdiagnostic happenstance likely occurred, because tuberculosis and AIDS have several notable correlations. First, tuberculosis was the "single most important HIV-related opportunistic infection in African countries." Second, tuberculosis (along with bacterial infections) was the main cause of **morbidity** (sickness) and **mortality** (death) among African AIDS patients. Third, in African AIDS patients, tuberculosis was correlated with AIDS-wasting ("weight loss").[344-347] Thus, an untreated tuberculosis patient having three symptoms belonging to tuberculosis; namely, weight loss, a persistent cough of 1 month's duration, and a persistent fever of 1 month's duration – could have been "diagnosed" with AIDS, per the WHO provisional surveillance definition. The populations in which the surveillance definition was "validated" contained many patients whose most common presenting symptoms were weight loss, chronic diarrhea, chronic cough, and fever as presenting symptoms.[339, 341] In one study, tuberculosis was the dominant pathological finding in seropositive patients who died from severe wasting.[348] Other diagnostically confounding medical conditions included diabetes, diarrhea with weight loss, and renal failure.[339, 345]

In 1991, a graphically critical editorial review was published in the *British Medical Journal*.[345] The author criticized the surveillance definition as an "unworkable concept." First, the clinical diagnosis of AIDS *did not require* any specific diagnosis of the underlying cause of the clinical symptom. Second, tuberculosis was the primary AIDS-related diagnosis in Africa, and pulmonary tuberculosis was NOT an AIDS-defining condition at that time. Third, the validation of the definition had been insufficient due to the use of HIV seropositivity as the gold standard of measure (whereas the surveillance definition of the U.S. Centers for Disease Control AIDS would have been the preferred measure). Last, the definition was proposed as a surveillance tool, but it was being applied for diagnosis and patient management. Clinically, the provisional definition was "unreliable." Also, tuberculosis met the surveillance definition criteria for AIDS, and could be wrongly classified as AIDS.[345]

Moreover, citing several studies, this author noted that: (1) the positive predictive value of the provisional definition dropped to 30% amongst confirmed AIDS patients; and (2) much HIV-related morbidity and mortality were not due to AIDS-defining conditions, but rather as result of non-opportunistic bacterial infection. He concluded that the clinical effect of the surveillance definition was to place undue emphasis on opportunistic infections and AIDS, to the exclusion of other problems.[345]

Another author, borrowing data, calculated that the surveillance definition had a false-negative rate of 60%; that is, the surveillance definition did not diagnose AIDS in 60% of AIDS patients. The author concluded that AIDS diagnoses were being assembled out of clinical features but such diagnoses were excluding "locally more sensitive diseases like tuberculosis."[349] (The word "sensitive" here is being used in the sense of an assay – sensitive to detection.)

Taken together, the comments of these two authors suggest that the provisional surveillance definition does not necessarily "see" HIV infection (AIDS).[349-351] Rather, what the surveillance definition "sees" is seropositivity, and this seropositivity had been designated as HIV infection (a true finding in some instances, no doubt).[*]

In a parallel thought, another commenter wrote: "Such [diagnostic] difficulties are amplified where morbidity not associated with AIDS, caused by non-opportunistic bacteria and *Mycobacteria*, is common."[345] **Mycobacteria** are bacteria with some fungus-like qualities, the tuberculosis germ being one.

Other authors shared a few viewpoints with this author and suggested modifications.[339, 342, 350, 352, 353] In 1994, the definition was expanded, adding several clinical diseases and conditions.[354]

---

[*] Author's conclusion

Chapter 14

African versus Classical AIDS

In Africa, the clinical presentation of AIDS differed from that observed in the United States and Europe. For the purposes of this discussion, the manner in which AIDS manifests within American and European patients will be referred to as "classical" AIDS."

Clinical Presentation

In Africa, the most common symptoms among patients purported to have "AIDS" were profound weight loss and severe chronic diarrhea (the second likely inducing the first) along with fever and cough. The most common early onset symptoms were also profound weight loss and severe chronic diarrhea. Tuberculosis was also a pathological finding associated with AIDS-wasting ("weight loss").[339, 341, 344, 345, 351, 353, 355-357]

In classical AIDS, the most common early onset symptoms were *Pneumocystis carinii pneumonia* (PCP), Kaposi's sarcoma (KS), and oral candidiasis (a type of yeast infection that can occur with or without HIV infection). In classical AIDS, appropriately, the first blossoming opportunistic infections of HIV infection were viral, fungal, protozoal infections, and tumors. These conditions arise during HIV infection because HIV kills off the T4-cell population, and the T4-cell triggers the immunological system into action against these conditions (cell-mediated immunity). Yet in African AIDS patients, these viral, fungal, and protozoal infections were far less prevalent. In some African AIDS patient settings, bacterial infections were responsible for much of the morbidity and mortality.[340, 344, 358-360] But bacterial infections should not be the first to appear in AIDS patients, since the T4-cell is not required for immunological response to fight off bacteria (humoral immunity). Rather, humoral immunity should remain intact and continue to fight off bacterial infections until the late stages of HIV infection, essentially when the body is "falling apart."

Oddly, both KS and PCP were far rarer in African AIDS patients than in their European or American counterparts.[214, 344, 345, 347, 355, 357, 361-367] The absence of KS should have been contrary to expectations. After all, in the first medical and media broadcast of AIDS, Kaposi's sarcoma was reported as endemic in equatorial Africa, the same region purportedly the epicenter of African AIDS. The absence of PCP was also nonsensical, and should have raised a red flag, since the causal agent *Pneumocystis carinii* is reportedly ubiquitous worldwide.[368, 369] [The historical absence of diagnosis is perhaps due to the lack of diagnostic facilities.] One author cites a serological survey of Britain and Gambia showing 70% seroprevalence of antibodies to *Pneumocystis carinii* by eight years of age in both countries, similar to levels observed in the United States.[370-372]

More recently, PCP has been reported among the HIV-infected in African populations,[373] particularly among HIV-infected infants.[374] However, the individual studies among these reports contain a number of patients, and the prevalence of tuberculosis and/or bacterial pneumonias usually exceeds that of PCP.[346, 369, 375-379] Also, in a review that listed PCP prevalence rates among cohorts in Burundi, Tanzania, Côte d'Ivoire, Botswana, and South Africa; only South Africa had a PCP prevalence rate (22%) comparable to that of American AIDS patients, whereas tuberculosis rates ranged from 28–75%.[346]

As seen in the United States, PCP was a swift killer. Before the development of effective PCP **prophylaxis**, it was the deadliest presenting symptom and the primary cause of AIDS-related mortality. The mortality rate of PCP, as a presenting symptom, was double that of KS. Among the first 159 patients identified by the CDC, 40% percent died of the initial PCP infection.[7, 8, 380]

African AIDS patients also frequently had itchy, hive-like rashes characterized by macules (spots) and papules (solid elevations of skin) on their arms and legs, not something common in classical AIDS but a trait seemingly shared with Haitian AIDS patients.[344, 381, 382] Female African AIDS patients also experienced a condition never seen before in AIDS patients: amenorrhea – the cessation of menstruation.[58, 383] The development of amenorrhea might be explained by profound weight loss, similar to the anorexic phenomenon among ballerinas, described previously.

In the United States, opportunistic Kaposi's sarcoma secondary to HIV infection was, almost exclusively, a homosexual phenomenon, due to the sexual transmission of an infectious KS agent: the Kaposi's sarcoma herpes virus (KSHV) a.k.a. human herpes virus 8 (HHV-8), evidently transmitted by fecal contact.[27, 384-396] The prevalence of Kaposi's sarcoma was far lower among the other American risk groups.[11, 25]

AUTHOR QUESTION

Tuberculosis resurged in the United States and other developed countries between 1985 and 1992. This increase has been habitually attributed to the AIDS epidemic. (In the United States, the populations for HIV infection and tuberculosis overlap). Other recognized contributing factors were the increased prevalence of homelessness, a deteriorated public health infrastructure, and the development of multi-drug resistant tuberculosis.[397] Drug resistance among strains of *Mycobacterium tuberculosis* and *Plasmodium falciparum* (malaria) has been long noted. The Author questions whether analogous drug resistance might have developed among other endemic pathogens, such as *Cryptosporidium species*? Or have other alternations in the pathogen-host equilibrium, such as factors affecting the exposure or susceptibility of the host, been culprit? Furthermore, given the global disposure of industrial toxins, might any lymphotoxic or **cytotoxic** (cell-killing) toxins reside in the environment(s).

Theoretically, it is possible that the profound weight loss and chronic diarrhea common to African "AIDS" patients are a consequence of viral, fungal, protozoal infections, and tumors secondary to HIV infection. Yet, other parts of the puzzle are missing, such as the absence of KS and PCP; never mind the implausibly high rates of heterosexual HIV transmission purported to occur among African heterosexual populations (discussed in *"HIV Transmission"*). Also, protozoal infections leading to death occur in absence of HIV infection, particularly in the presence of malnutrition.[366] (The effects of malnutrition on the cell-mediated immunity is discussed in *"Other Forms of Immunodeficiency."*)

<u>Aggressive Kaposi's sarcoma</u>

In Africa, there were reports of a new aggressive form of Kaposi's sarcoma that was differentiated from the historical endemic form. The emergence of this new aggressive Kaposi's sarcoma was reported in 1983 in Lusaka, Zambia.

In 1983, thirteen (13) patients presented with an aggressive form of KS that was terminal for 8 patients within the year while 10 other patients presented with endemic KS (cutaneous, slow-progressive, responsive to therapy). The patients with the aggressive disease were younger, better educated, and had better jobs; four of them were professionals of the same background, suggesting an occupational or social cluster. None had visited the US. Only one had traveled extensively across Europe, and only two men admitted to single episodes of homosexual behavior.[398]

Northeastern Zaire also seemed to experience an outbreak of Kaposi's sarcoma. The yearly prevalence rate was 7 cases a year before 1983. In 1984 and 1985, the prevalence rate had jumped to 14 cases a year. (At the end of 1984, the United States had 3600 AIDS cases.[9] At the end of 1985, the United States had 15,179 cases.[10]) No relevant clinical or serological data was reported on this cohort.[399]

In both of these reports, the reporting doctors naturally compared this manifestation of KS with that observed in American gay men and AIDS. This new form of aggressive Kaposi's sarcoma seemed to follow on the footsteps of the American and European AIDS epidemic. HIV antibody testing was not yet available at the time of these reports. Later reports would correlate the aggressive KS form with HIV seropositivity.[214, 219] However, a third author mentioned that a generalized, aggressive Kaposi's sarcoma, distinct from the endemic form, had been described in Uganda since at least 1962 (without citing his sources).[355]

ENVIRONMENTAL KS FACTOR?

Historical outbreaks of Kaposi's sarcoma might be viewed in a different light with the understanding that the manifestation of Kaposi's sarcoma is a consequence KSHV infection (HHV-8) in an immunocompromised host (although, theoretically, high concentrations of exposure to a pathogen could overcome the immune system). In Africa, KSHV transmission frequently occurs in childhood via saliva transmission.[400, 401] Exposure to iron has also been suggested as a contributing factor to KS development (iron being oncogenic), perhaps among populations exposed to iron-rich volcanic soils.[402, 403]

Slim Disease

"Slim" disease was first reported from Uganda as a "new disease" in 1983, seemingly appearing contemporaneously with the American, European, and Central African AIDS epidemics. (It was also referred to as **enteropathic** AIDS, enteropathic meaning pathology of the intestine.) Overall, slim disease was characterized by profound weight loss and chronic diarrhea, but its character varied in early reports. (Chronic diarrhea can be deadly; the patient dies of dehydration and/or electrolyte imbalance.)

Per the authors of the first report: " 'Slim' disease is a new syndrome hitherto unreported in Uganda. It is clearly associated with HTLV-III infection and is not unlike AIDS. It is most unlikely that the disease has not been reported before 1982, since medical records in Uganda are good and go back to 1944."[355] However, lymphadenopathy and Kaposi's sarcoma were not associated with slim disease, "although KS is endemic in this area of Uganda."[355]

Of the 42 slim disease patients, only 4 had KS, one with disseminated disease, and three had cryptococcal meningitis, a common AIDS-related, opportunistic infection. Thirty-four had oral candidiasis and the same number was HIV seropositive (ten percent of the healthy hospital staff also tested seropositive). Out of a combined population of 71 slim disease patients, the male-to-female ratio was 1.2:1. Mortality information was not provided, which is interesting in that the mortality rate of AIDS was such a striking feature.[355] This study is the most frequently cited article on slim disease in Africa.[355]

One year later, another Ugandan report described 23 **enteropathic AIDS** patients, including 7 females. Enteropathic AIDS was characterized by oral thrush, diarrhea and weight loss. Twenty-two of the 23 patients had oral candidiasis; and 1 patient had possible *Mycobacterium avium* mycobacteriosis, an AIDS-defining condition. Eleven patients had cryptosporidiosis (*Cryptosporidium*) and three had isosporiasis (*Isospora belli*), both opportunistic protozoal enteric infections common to AIDS patients. Protozoan infections are hallmarks of HIV infection, although such infections can occur in the absence of HIV infection. (In Denmark, only seven of the first 231 AIDS patients had cryptosporidiosis.[404]) Mortality rates were not provided. Per the report authors: "Twenty-three typical patients were selected, *none of whom appeared to be near to death.** Only those patients with the triad of diarrhea, weight loss and oral thrush were studied, and we believe that this subgroup is representative of the enteropathic form of AIDS in Uganda."[405]

A third author reported on ten slim disease patients: 8 women and 2 men. "All had marked weight loss with extreme fatigue, marked diffuse wasting with significantly decreased circumferences of arms, thighs and calves." Seven of the ten afflicted had chronic diarrhea and chronic fever. Five had pruritic dermatitis. Three had psychomotor slowing. Two had lymph node tuberculosis. One had generalized lymphadenopathy. One had oral candidiasis. The main finding of the muscle biopsy was atrophy of muscle fibers.[406] Oddly, only five of the ten patients with slim disease met the CDC criteria for "HIV wasting syndrome," defined as

---

* *italics* added

involuntary weight loss (> 10% of baseline body weight) plus either chronic diarrhea (for 30 days or longer) or chronic weakness and documented fever (for 30 days or longer) in the absence of a concurrent illness or condition other than HIV infection that could explain the findings.[406, 407]

Overall, every one of these authors stated that these disease outbreaks were highly suggestive of AIDS. In apparent response, the WHO surveillance definition was seemingly designed to incorporate the characteristics of slim disease into the provisional surveillance definition.[286, 408] "We now have a provisional WHO clinical definition, which can more easily accommodate 'slim' disease," reports the head of the AIDS program at WHO as quoted by *Lancet* in December 1986.[409]

In the United States, extreme weight loss was not a typical, early onset presentation of AIDS. All conditions are possible, but *profound* weight loss was not a frequent early presentation event. Weight loss was a frequent component of lymphadenopathy and/or AIDS prodrome, but not marked or profound weight loss; the AIDS prodrome, by its very nature, had a persistent vagueness. Initially, the presence of the AIDS prodrome was usually recognized only in hindsight, usually after the manifestation of some more obvious opportunistic disease.

An AIDS wasting syndrome is recognized as occurring in the United States, but it only received an honorable mention, rarely the featured key point of any published reports on AIDS / HIV infection. AIDS-wasting in African patients also differs from the **lipodystrophy syndrome** characteristic of classical AIDS patients on multi-antiviral drug regimens. In such AIDS patients, lipodystrophy is a disorder characterized by abnormal bodily fat distribution.

HIV Transmission

In a realistic consideration of HIV transmission in Africa, one must question the purported efficacy of transmission by the African heterosexual population. An early African study of 150 couples demonstrated that both members were seropositive in 90% of couples;

illustrating, say the authors, "the high efficiency of heterosexual [HIV] transmission" in this African population.[410] Such high rates of HIV transmission are nonsensical when compared with heterosexual transmission rates of American and European heterosexual couples.

Comparatively, far different rates of HIV transmission were determined in two California studies conducted 10 years apart. In 1987 and 1997, **male-to-female transmission** among discordant couples (one partner seropositive; one partner seronegative) was 20% (61/307) and 19% (68/360), respectively. **Female-to-male transmission** was 1% (1/72) and 2.4% (2/82), respectively.[411, 412]

Among Europeans – 563 discordant couples from 13 medical centers – an overall 19% of men and 12% of women contracted HIV.[413]

In Italy, among 730 discordant couples, the transmission rates were 24% male-to-female and 10% female-to-male.[414] In a meta-analysis of 16 studies of similar design, one author reported the following overall transmission rates tabulated below (with the caveat that transmission rates varied widely, ranging from 1% to 73% in female-to-male transmission, and 16% to 61% in male-to-female transmission):[415]

United States or Europe	Positive	*n*	Transmission
Male-to-female transmission	411	1485	28%
Female-to-male transmission	67	461	15%
Africa or Haiti	**Positive**	***n***	**Transmission**
Male-to-female transmission	171	324	53%
Female-to-male transmission	83	143	58%

Missing Infections

Pneumocystis carinii pneumonia (PCP)

As previously stated, PCP is notably lacking among African AIDS patients. One astute author noted the lack of PCP as well as the almost total absence of two other diseases endemic in various African locales. In 1990, he wrote: "One of the most striking clinical differences between AIDS in developed and developing countries is the low incidence of PCP in Africa. 65% or more of homosexual AIDS patients in Europe and North America develop PCP, whilst its prevalence in small series of African AIDS patients being treated in Europe is only 28.%"[366]

This author also noted a total absence of PCP among purported AIDS cases among some Zairean and Rwandan cohorts,[57, 58, 366] and its only occasional appearance among other Zambian and other Rwandan cohorts. PCP was present among 22% of Zimbabwean patients (8/37). Yet, in Uganda, no PCP was found in 22 AIDS autopsies. Also, broncho-alveolar lavage (a process by which cells and fluids are removed from the bronchi and lungs) of 40 Zambian patients found no PCP. [Many of these patients and cohorts are also described in this book.]

Strongyloidiasis

Strongyloidiasis is a disease caused by *Strongyloides stercoralis*, a nematode. **Nematodes** are known as roundworms. Humans become infected by ingesting eggs in contaminated food or by transmitting eggs to the mouth by the hands after contact with contaminated soil.

In severe cases, strongyloidiasis induces a hyperinfection of the intestine with or without extra-intestinal dissemination with moderate to severe diarrhea, and possibly pneumonia and CNS infections.

Other groups known for having impaired cell-mediated immunity and susceptibility to strongyloidiasis include organ transplants recipients; and leukemia and lymphoma patients (including T cell types). The majority of these patients and those with other underlying conditions have also received steroid chemotherapy, systemic corticosteroid therapy of duration greater than 1 month is known to be immunosuppressive themselves.[366]

"Whilst some 15% of patients with severe strongyloidiasis have no definable defects in cell-mediated immunity, the implication of the majority of cases is clear: depression of cell-mediated immunity by HIV infection should result in severe strongyloidiasis wherever that parasite is endemic."[198, 366] In 1986, the CDC/WHO surveillance definitions included strongyloidiasis as what is now called an "AIDS-defining condition" (in conjunction with HIV seropositivity and/or other suitable laboratory and/or clinical manifestations). Yet by the time of this report (1990), only 4 cases of HIV-related strongyloidiasis had been reported in the U.S. One had been reported in Columbia, and one in Zaire.[366] In Africa, only small percentages of seropositive populations had strongyloidiasis: 1/42 patients in Zaire and 3/44 patients in Zambia, even though the parasite has prevalences of 25 – 48% in central African populations. None of the symptoms nor histology or these two groups indicated the presence of any hyperinfections. [366] The 1987 Revision of the CDC Surveillance Case Definition eliminated disseminated strongyloidiasis as a criterion.[416]

The author also concludes that "slim disease" is not a consequence of strongyloidiasis; rather, the etiology of "slim disease" is multi-factorial ("slim disease" is discussed in the *"Slim Disease"* sub-section of the chapter *"African versus Classical AIDS"*).

Amoebiasis

Amoebiasis is a disease caused by *Entamoeba histolytica*, an amoeba. *Entamoeba histolytica* infections occur worldwide in areas of poor sanitation. *Entamoeba histolytic* is excreted by infected people, and is transmitted to people who ingest contaminated food or water (oral-fecal route). Amoebae infect the large intestine; possibly inducing inflammation and ulcers. They may also infect the liver and induce abscesses. Symptoms include severe weight loss, nausea, vomiting and diarrhea.

Entamoeba histolytica is a protozoan. T4-cell defect induced by HIV infection causes a failure in the triggering mechanism of cell-mediated immunity (the T4-cell being the trigger), so viral, fungal, and protozoal infections have the opportunity to grow unchecked. In cases of legitimate HIV infection in locales wherein *Entamoeba histolytica* is endemic, one would

expect a high prevalence of rampant amoebiasis on the PLWH.

In the U.S., *Entamoeba histolytica* seems to be relatively common among gay men but

without deleterious effects.[366, 417] Also, a low level of mortality among the HIV-infected could

be assumed from the relative lack of medical literature regarding describing the pathology

and/or prevalence of *Entamoeba histolytica* among HIV/AIDS patients. The literature is

virtually absent.

Nevertheless, the author states: "Classically, invasive amoebiasis is increased in

incidence and severity by factors that depress cell-mediated immunity, i.e. steroid therapy,

pregnancy and malnutrition . . . . In Africa, where pathogenic strains of *Entamoeba histolytica*

are common, the lack of clinical or pathological evidence of invasive amoebic [infections] in

symptomatic HIV-infected people is striking. There are insufficient data yet to indicate the

converse – a protective effect of HIV against invasive amoebiasis . . . . Thus it is surprising that

HIV infection of [T4] lymphocytes and macrophages does not impair host defenses against *E.*

histolytica." [366]

At outset, the author states: "Serious infections with *Entamoeba histolytica* and

Strongyloidiasis stercoralis would be expected to occur in AIDS patients: they do not." The

author does not consider that some African populations under study likely do not have

legitimate HIV infections. The author's deliberations were made in the belief that HIV was a

burgeoning epidemic in Africa. The author stressed the importance of the pathogen-host

interaction in the gut mucosa (mucous membranes) in susceptibility or defense against

infection; conjecturing that host defenses at this site might persist despite HIV infection or that

HIV infection somehow enhances host intestinal defense capability is these specific enteric

agents.

The author suggest that the lack of PCP mortality in some parts of central Africa might be due to patients dying of more virulent infections before they acquire PCP.[366]

Both *Entamoeba histolytica* and *Strongyloidiasis stercoralis* are capable of causing death, but neither seems prominent in the medical literature as a cause of mortality during HIV infection in the Western world, despite *Entamoeba histolytica* (the amoeba) being endemic among the gay male population.

A hospital-based French group also cited the lack of strongyloidiasis, and noted that most of the patients with strongyloidiasis had been on high-dose corticosteroids; whereas, "HIV infection seems to favor coccidian rather than the helminthic infestations."[418] *Coccodia* are a subclass of single-celled protozoans, several of which are common AIDS-related opportunistic infection. The **Coccodia** contain *Cryptosporidium* (gastrointestinal illness with diarrhea), *Toxoplasma gondii* (frequent cerebral infections; house cats are carriers); and *Isospora belli* which infects the small intestine. Helminths are parasitic worms, such as *Strongyloidiasis stercoralis*, that tend to live in the intestines of animals. It is possible that the specificity of the T4-cell defect allows viral, fungal, protozoal opportunistic infections, the protozoal component specific to coccodian infections. (Or, the misdiagnosis of other diseases in Africa as HIV/AIDS has adversely impaired such scientific deduction.)

Kaposi's sarcoma

The overall lack of KS in the African populations is striking. In its first report of KS manifestations among gay men in the United States, the CDC cited a report on the prevalence throughout black and white populations in Africa published in 1972: Kaposi's sarcoma was "endemic belt across equatorial Africa, where KS commonly affects children and young adults and accounts for up to 9% of all cancers."[419] Central Africa includes Zaire (now the Congo), the purported epicenter of the HIV/AIDS epidemic.

A single sero-prevalence study performed in the Republic of South Africa is also particularly striking. In 2008, 2103 South Africans (95 sex workers, 862 miners, and 731 female and 415 male town residents) were tested for HIV, KSHV (Kaposi's sarcoma herpes virus), and several sexual transmitted diseases.[420] (Please see Chapter 14 for a discussion of KSHV).

Overall, KSHV and HIV prevalences were 47.5% and 40%, respectively ($p = 0.43$; this high p value means that these percentages were not statistically different or valid).[420] In the United States, opportunistic Kaposi's sarcoma secondary to HIV infection was, almost exclusively, a homosexual phenomenon, due to the sexual transmission of an infectious KS agent: the Kaposi's sarcoma herpes virus (KSHV) a.k.a. human herpes virus 8 (HHV-8), evidently transmitted by fecal contact.[27, 384-396] Thus, the manifestation of Kaposi's sarcoma is induced by a viral infection within immunocompromised people.

Thus, given legitimate HIV infections in the Republic of South Africa, one would Kaposi's sarcoma to be commonplace among purported HIV/AIDS patients. Given KSHV and HIV prevalences of 47.5% and 40%, respectively, either these patients lack legitimate HIV infections and/or the KSHV outcomes are also questionable.

Interestingly, in this South African population, KSHV was not statistically associated with any sexually transmitted infections, HIV, or sexual behaviors, i.e., the analyses indicate that KSHV was not transmitted sexually in this population.

Table 17 – *New York Times* Headlines about HIV/AIDS in Africa

"AIDS in Africa: A Killer Rages On"

New York Times (Front Page), September 16, 1990

"Parts of Africa Showing H.I.V. In 1 in 4 Adults"

New York Times (Front Page), June 24, 1998

"In Zambia, the Abandoned Generation"

New York Times, September 18, 1998

"Doctors Powerless as AIDS Rakes Africa"

New York Times (Front Page), World: August 6, 1998

"AIDS Is Everywhere, but Africa Looks Away"

New York Times (Front Page), December 4, 1998

"Faulty Condoms Thwart AIDS Fight in Africa"

New York Times (Front Page), December 27, 1998

"H.I.V. Is Linked To a Subspecies of Chimpanzee"

New York Times (Front Page), February 1, 1999

Chapter 15

Outliers – The Exception as the Rule

By a bevy of accounts, AIDS existed in Africa decades before it appeared in NYC. Supposedly, between 1959 and 1981, twenty or so individual AIDS cases and a couple of "epidemics" emanated out of or occurred in Africa.[421-423] These AIDS cases are outliers.

Technically, an **outlier** is an observation that differs so widely from all others in a set that one must conclude that a gross error has occurred; or the outlier comes from a different population. In this discussion, outliers represent the singular, disparate cases of AIDS removed in time and geography from all other documented cases. The "clusters" of HIV/AIDS discovered in the time period were done so by retrospective serum or tissue analysis: fishing expeditions using ELISA or some other immunoassay.

By the Author's definition, they distinctly are not *reasonable* AIDS cases: they do not appear in infectious disease clusters and, upon close examination, the reported symptoms rarely match those of classical AIDS. By offering these outliers as proof that AIDS existed in Africa prior to 1981, the proponents of this theory use the exception to prove the rule, a basic scientific *faux pas*.

Prominent Outliers

In one prominent and widely publicized case, a seaman from Manchester, England reportedly had AIDS in 1959. He had been diagnosed with *Pneumocystis carinii* infection and the case had stumped his doctors. In 1983, an attending physician wrote a letter to the scientific journal *Lancet* entitled: "AIDS in 1959?" He described a clinical profile matching the AIDS prodrome and the autopsy finding of PCP. The patient had been in the Navy between 1995–1957 and "had traveled abroad (the presumption being Africa?)."[424]

In 1990, scientists reportedly extracted HIV from the seamen's preserved tissue. In the end, this "discovery" of HIV from 1959 was actually another incident of laboratory contamination.[424-431] It was also later clarified that although the English sailor in the Royal Navy, but "was stationed exclusively in England, except for a brief voyage to Gibraltar in 1957."[429] This patient apparently had some form of cell-mediated immunodeficiency, but it was unlikely to have been HIV infection. As usual, the retraction gained far less notoriety than the initial false finding. Today, the belief in the validity of this disproved AIDS case stubbornly persists among the lay and professional communities alike.

Another prominent case is that of a Danish surgeon, a previously healthy 47-year old woman who died in 1977. She had worked as a surgeon at a primitive, rural hospital in Zaire in 1972–1975. In 1976, she presented with diarrhea, fatigue, wasting, and lymphadenopathy; the conditions resolved but a year later she developed PCP and oral candidiasis.[49] The assumption is that she contracted HIV by being exposed to blood and/or bodily substances during surgery under these conditions. Very likely, she had some form of impaired cellular immunity, but the timing and placement of this condition makes the presence of HIV unlikely.

"Puzzle of Sailor's Death Solved After 31 Years: The Answer Is AIDS"

The *New York Times*, July 24, 1990

A Norwegian family – father, mother, and daughter – is frequently cited as the first AIDS cluster. The husband was a seaman who had traveled along the West African coastline in 1962–1965. In 1966, at the age of 20 years, he presented with muscle and joint pain, a skin disease reminiscent of measles (maculopapular exanthema), lymphadenopathy, and recurrent respiratory infections. He remained stable until 1975 when his condition deteriorated because of pulmonary disease and neurological manifestations (ascending paralysis of the legs, loss of muscle coordination, incontinence, and dementia). He died in 1976.[432, 433]

This case is not reminiscent of AIDS in that it lacks any presentation of viral, fungal, or protozoal infections and/or malignancies; although, theoretically, his neurological conditions could have resulted from HIV infection. (HIV is neurotropic: it infects neurons in the brain. Other opportunistic diseases can also infect the central nervous system to induce such symptoms.) Also, a ten-year survival period in absence of effective anti-retroviral treatment – nonexistent at the time – is highly unlikely.

HIV HAS A 10-YEAR INCUBATION PERIOD - NOT

Contrary to popular belief, HIV does not necessarily have a 10-year incubation period. Incubation is the period from initial infection until the onset of opportunistic diseases (such as PCP, KS, or candidiasis). Contrary to popular belief, HIV does not necessarily have a 10-year incubation period.

By March 1983, even before HIV was discovered, the CDC estimated that HIV infection had an 8- to 18 month incubation period.[434] AIDS was a mystery at the outset, and health authorities had actively tracked and interviewed patients when possible to gather information both about possible etiology and transmission vectors (both unknown at that time). Not much of this empirical data was published in the medical literature and cohorts reported by CDC are very small, but in each cohort, a portion of the patients had short incubation periods.

As stated by the CDC: "Furthermore, the California cluster investigation and other epidemiologic findings suggest a 'latent period' of several months to 2 years between exposure and recognizable clinical illness and imply that transmissibility may precede recognizable illness."[47] (Among infants born with HIV infection, the incubation period was typically 6 months or less, although a "failure to thrive" was sometimes immediately evident.[47, 435-438] As of 1988, the median survival of infants born with HIV infection, or infected with HIV by transfusions, was 8 – 9 months.[439])

The "California cluster" referred to by the CDC was a cluster of 19 gay men in California with PCP and/or KS: 7 alive and 11 dead. Of the 19 patients, sexual partner data was obtained for 13 patients. Within 5 years of symptom onset, 9 of these patients had sexual contact with other KS of PCP patients. Three of the 6 patients with KS developed their symptoms after sexual contact with persons who already had symptoms of KS. One of these 3 patients developed symptoms of KS 9 months after sexual contact. Another patient developed symptoms 13 months after contact, and a third patient developed symptoms 22 months after contact.[440] This evidence on incubation from this cohort of 19 is less conclusive than the Danish study, described below. It seems likely that these three patients, or any member of this cohort, may also have had one or more sexual contacts with men who were infected with HIV and **viremic** (the presence of viruses in the blood), but were asymptomatic.

In one published study, the CDC had tracked 40 AIDS patients in 10 cities linked by sexual contact. Based on 6 patients in this cohort, as determined by direct patient interview, the CDC estimated a mean latency period of 10.5 months. In these 6 patients, symptoms were first noticed a mean of 10.5 months (range 7 to 14 months) between sexual contact and symptom onset. The sexual contact was with one of four cohort members who also was a reported patient.[441]

Similar but more definitive findings were found in Denmark. In Denmark, AIDS first appeared among a cluster of homosexual men who had traveled to the United States. Eight of the first 20 AIDS patients in Denmark had visited the United States, which was then recognized as the epicenter of AIDS (New York, San Francisco, and Los Angeles, more precisely). To quote the authors: "Assuming that the travelers did contract the disease while in the USA, speculations about the incubation period might be relevant. The median period between departure from the USA and the diagnosis was 12 months (range 4 – 39) and the median period between arrival in the USA and the diagnosis was 25 months (range 6 – 41). Although our data indicate an average incubation period of 1-2 years, we cannot rule out the possibility that it may be even longer." These Danish authors also reported a 6-month incubation period. The subject presented with opportunistic infection 6 months after sexual contact in the United States.[36]

A few sporadic, shorter incubation periods have been reported:

- a 2-month incubation period following homosexual male contact;[442]

- a 2-month incubation period from intravenous needle transmission to the

  development of lymphadenopathy;[443]

- a 7-week incubation period from a blood transfusion to the development of

  AIDS-related complex;[444] and

- a 6-month 7-week incubation period following a blood transfusion.[445]

One author reported two short-term period of progression from "primary

infection" to AIDS. A 32-year old homosexual male presented with a 4-day fever,

abdominal pain, and diarrhea with blood: what the author seems to assume was an

acute reaction to an HIV infection that occurred some weeks prior (the author indicates

he knows the time frame of exposure but does not describe it). In two weeks he

developed lymphadenopathy, and within 2 months of initial fever, he developed PCP

(along with seroconversion to HIV positive).[445]

The same author reported a 73-woman who received 1 unit of blood. Two

weeks after transfusion, she had acute illness with fever, malaise, weight loss, and

lymphadenopathy which resolved without treatment after 8 weeks. She developed AIDS

(seropositive plus cerebral toxoplasmosis) within 6 months after blood transfusion.[445]

These findings do not necessarily countermand the existence of individuals, or

some portion of the HIV-infected population, that survive for 11 years without treatment

or experience a silent 10-year incubation period. Yet in these virgin, untreated groups,

6 patients (among a network of 40 sexual contacts) developed AIDS (manifested

opportunistic infections) in a median 7.5 months after exposure (range 7 to 14 months),

and 8/20 AIDS patients had average incubation periods of 1-2 years (range 4 to 39

months).[36, 434] The numbers in these available reports are very small, yet the incubation

period they document likely represents the beginning edge of the distribution curve for

HIV incubation.

Moreover, these first few cases heralded an incipient tsunami. For the first two

years, the AIDS numbers in the United States doubled very 6 months.[7] By 1985 it had

slowed to doubling every 12 months or so.[446, 447] By 1987 or so, the wave had crested, such that the annual prevalence increased only 4.9% from 1990 to 1991 (though the momentum of the threshold population continued to induce accumulation).[448]

Yet, in this time, humans were undergone a herd behavioral change. The gay bath houses closed and the culture dissipated; the gay discos had become Russian Roulette, and needle interventions were enacted via needle exchange and/or bleach. (And possibly, a significant proportion of those people at greatest risk had already died).

Also, during this time: the etiology of AIDS remained unsolved for 4 years • the CDC developed and distributed a surveillance definition based on utilizing clinical and immunological profiles • HIV was discovered and identified as the etiological agent • an HIV testing kit was developed and approved • modified CDC surveillance definitions were developed and distributed along with the establishment of an infrastructure for the purpose of distributing information to both clinicians and the general public • HIV/AIDS became mainstream news • a first-time ever human anti-retroviral drug was developed • and a radical political advocacy group named "ACT-UP!" was formed. All these events occurred in less time than the purported median longevity of any particular HIV/AIDS patient, if one considers a 9 – 11 years (in developed countries) as median.

The mother of the Norwegian family had myelogenous leukemia (involving white blood cells but not lymphocytes), candidiasis, cytomegalovirus, lymphocyte depletion, and a progressive neurological condition consequent to an encephalomyelitis. (**Encephalomyelitis** is inflammation of the brain and spinal cord; **cytomegalovirus** is a viral infection common to highly sexually active, gay men with and without HIV-infection.) After birth, the daughter developed normally for 2 years, and then suffered from recurrent bacterial infections. Her condition was stable for many years before she died of another viral disease common to AIDS patients. Also, all family members also had **hypergammaglobulinemia** (an excess of antibodies), an abnormality of humoral immunity. This condition is not necessarily an indication of HIV infection. HIV disrupts cell-mediated immunity. Humoral defects can also occur during HIV infection.

Overall, this family cluster suggests some form of immunological disorder possibly transmitted sexually from father to mother, and "vertically" from mother to child, but other explanations are possible. For example, the father and daughter could have had genetic immunological disorders, while the mother simply had leukemia.[432]

Nonetheless, years later, investigators successfully isolated HIV viral DNA from preserved family autopsy tissues (amplified and reconstructed segments). Naturally, this HIV viral DNA recovered from these autopsy segments belongs to HIV Group O (outlier).[449] (HIV is a retrovirus, using RNA as the blueprint for life; but once HIV infects a host cell, HIV "reverse-transcribes" viral DNA as an intermediary step in the replication process.)

"Norwegian Family Got AIDS in 1960s, Researchers Report"

Chicago Tribune, June 21, 1988

In 1987, viable HIV was also reportedly isolated from frozen Central African serum samples stored since 1976.[450] This finding apparently escaped notice within the scientific community. Also, nucleotide sequences of HIV viral RNA were also extracted from preserved tissue and sera samples dating back to 1959 from Léopoldville, Belgian Congo.[*] All extractions purportedly belong to HIV Group M (major) and all extractions involved PCR amplification of isolated fragments, one of these findings was reportedly confirmed by two separate laboratories.[247, 451, 452] These latter findings are more frequently cited in the medical literature. All these virological investigations generally begin with the finding of HIV seropositivity within the samples in question.

Only a very few intrepid researchers have searched the medical literature for evidence of HIV infection in populations outside of Africa. One literature reviewer found 19 cases

---

[*] now Kinshasa, *Zaire*; now Kinshasa, *Democratic Republic of the Congo*

suggestive of AIDS-like immunodeficiency in the United Sates and Europe between 1952 and 1979, as suggested by the manifestation of various infections and/or conditions common to HIV infection.[422] In the words of the reviewer:

"These cases retrospectively met the Centers for Disease Control's surveillance definition of the syndrome and had a clinical course suggestive of AIDS. The reports originated from North America, Western Europe, Africa, and the Middle East. The mean age of patients was 37 years, and the ratio of male to female patients was 1.7:1. Sixteen patients had opportunistic infections(s) without Kaposi's sarcoma. The remainder had disseminated Kaposi's sarcoma. The commonest opportunistic infection was *Pneumocystis carinii* pneumonia. Two patients were reported to be homosexual. Three others had been living in Africa, and one patient was born in Haiti. In two instances concurrent or subsequent opportunistic infection occurred in family members. All patients died 1 month to 6 years after the initial manifestation of disease. In view of the historical data, unrecognized cases of AIDS appear to have occurred sporadically in the pre-AIDS era."[422]

Table 17 lists the location, year of onset, patient age and /sex, the opportunistic infection(s) recorded, and the presence or absence of Kaposi's sarcoma in these 19 patients.[422] The authors conclude AIDS is an old disease that remained unrecognized in the past because of its sporadic occurrence.

Another set of reviewers, using aggressive Kaposi's sarcoma as a possible indicator of AIDS (cases wherein the patient died within 2.5 years of diagnosis), found 28 cases of aggressive KS in the United States and Europe dating between 1902 and 1966.[453] These 28 purported AIDS cases are listed in Table 18. These authors proposed that HIV infection is not new and did not come from Africa, but had been endemic in the Euro-American population at least since the beginning of this century. Also proposed is that sociocultural changes led to the spread of the virus, allowing for the epidemic to be recognized, and its subsequent introduction into Africa. The reviewer suggested that AIDS originated in the United States with a possible **zoonotic** origin in sheep, perhaps existing at undetectable endemicity — until sociological flux in 1970s changed the equilibrium to favor epidemicity. This sociological flux was concurrent with the tripling prevalence of both syphilis and Hepatitis B among the urban, gay populations during the same period.[453]

Table 17 lists the location, year of onset, patient age and /sex, the opportunistic infection(s) recorded, and the presence or absence of Kaposi's sarcoma in these 19 patients.[422]

Table 18 – Purported AIDS Cases in United States and Europe, 1952 – 1979

Location	Year Onset	Age/Sex	Opportunistic Infection(s)	Kaposi's Sarcoma
United States	1952	28/M	• Cytomegaloviral pneumonia	No
Canada	1958	36/M	• *Pneumocystis carinii* pneumonia	No
United Kingdom	1959	25/M	• Cytomegaloviral pneumonia • *Pneumocystis carinii* pneumonia	No
United States	1959	48/M	• *Pneumocystis carinii* pneumonia	No
United Kingdom	1961	12/F	• Disseminated *Mycobacterium avium* infection	No
United States	1964	22/F	• *Pneumocystis carinii* pneumonia	No
United States	1964	46/M	• Progressive multifocal leuko- encephalopathy	No
Canada	1966	40/M	• Progressive multifocal leuko- encephalopathy	No
Sweden	1967	23/F	• Disseminated *Mycobacterium kansasii* infection	No
United States	1968	15/M		Disseminated.
Israel	1969	59/M		Disseminated
United States	1969	56/F	• Progressive multifocal leuko- encephalopathy	No
Uganda	1973	45/M	• Disseminated strongyloidiasis	No
West Germany	1976	49/M		Disseminated
Denmark	1976	47/F	• *Pneumocystis carinii* pneumonia	No
Belgium	1977	34/F	• Disseminated cryptococcosis	No
United States, Hawaii	1978	52/F	• Intestinal cryptosporidiosis • Disseminated toxoplasmosis	No
West Germany	1978	21/M	• Disseminated *Mycobacterium fortuitum* infection	No
United States	1979	48/M	• Intestinal cryptosporidiosis • Disseminated cytomegaloviral infection	No

Disseminated refers to involvement of liver, bone marrow, or multiple organs.
Source: Table 2, *Unrecognized cases of AIDS in the pre-AIDS era*, in Huminer D, Rosenfeld JB, Pitlik SD. AIDS in the pre-AIDS era. Rev Infect Dis. Nov-Dec 1987;9(6):1102-1108.

Table 18 lists the 28 cases of aggressive KS found in the United States and Europe between 1902 to 1966.[453]

Table 19 – Aggressive, Disseminated KS, North America and Europe, 1902 – 1966

Age	Sex	Location	Disease Duration	Manifestations
58	M	Europe	2.5 years	Weakness, weight loss
38	M	Europe	1 year	Cachexia, marasmus
24	M	Europe	1.5 years	Weakness, vomiting, cachexia, ascites
51	M	Europe	1 year	Diarrhea, marasmus
44	M	Europe	0.5 year	
52	M	Europe	4 months	Cachexia
38	M	Europe	1.5 years	Disseminated tuberculosis
40	M	N/A	1 year	Hemolytic anemia, splenomegaly, leukopenia progressing to aplasia
26	M	Europe	2 years	
32	M	Europe	2 years	
56	M	Europe	0.5 year	
58	M	North America	2 years	Weakness, leukopenia, bloody diarrhea
50	M	North America	0.5 year	
33	F	N/A	2 years	Cough, fever, dyspnea
26	M	North America	2 months	Weakness, shortness of breath, night sweats, fever
46	M	South America	Autopsy	
46	M	North America	2.5 years	Weakness, fatigue
59	M	North America	1 year	Weakness, focal seizures secondary to Kaposi's sarcoma of the brain
59	M	Europe	1 year	
17	M	North America	Few weeks	
57	M	Europe	10 months	Lymphadenopathy with lymphoid hyperplasia, anorexia, melena (bloody stools)
43	M	South America	2 months	Dyspnea, weight loss, hemoptysis (spitting up blood), vomiting, anemia
32	M	North America	1 year	
27	M	North America	6 months	
24	M	North America	6 months	
44	M	Europe		
31	F	North America	2.5 years	
50	F	North America	4 months	

Source: Table 1. *Cases of disseminated Kaposi's sarcoma reported prior to aids epidemic* in Katner HP, Pankey GA. Evidence for a Euro-American origin of human immunodeficiency virus (HIV). J Natl Med Assoc. Oct 1987;79(10):1068-1072.

In closing, this second set of reviewers stated: "Therefore, it is proposed that the cases of disseminated Kaposi's sarcoma in the past were in fact manifestations of this virus, and that through sociocultural changes in society, not importation, this infection has become epidemic, and subsequently it was exported." [453]

The Author does not wish to contend whether or not these collections of AIDS-like cases are legitimate AIDS cases or not. Individual, case-by-case, clinical patient profiles were not provided by these reports. Collectively, these populations displayed a number of conditions indicative of impaired cell-mediated immunity. If thought to be AIDS cases: overall, they were singular, disparate cases lacking apparent risk factors or transmission vectors (excepting a couple of homosexual males in one report [422]). It is likely that some or all of these patients had some form of cell-mediated immunity, but not necessarily HIV infection. (Note: the author did not cross-reference these two collections to determine the extent of overlap.)

Looking back at the epidemiological history of HIV/AIDS, i.e., the burgeoning emergence of AIDS into the gay U.S. population, it seems likely that the introduction of HIV into the human domain was a sudden event. The emergence of AIDS was an abrupt onslaught, not the gradual emergence of a sub-endemic pathogenic entity. Specifically, AIDS emerged first almost exclusively among homosexual males, and not among the other primary risk group, IV drug users.

Among the first 107 AIDS cases in the United States (as of August 1981), 95% were white men 15-52 years of age. Ninety-four percent of the men for whom sexual preference was known (95/101) were homosexual/bisexual.[6] If HIV had progressed from sub-endemic to endemic to epidemic in the Western hemisphere, it seems likely it the emergence would have been a steadily growing trend among the populations at greatest risk for blood-borne diseases; namely, primarily highly sexually active homosexual males and IV drug users, and secondarily, the sexual partners of IV drug users and health care workers, the latter at risk via occupational exposure. Such was the profile of the "mature" endemic epidemiological profile of Hepatitis B virus (HBV) in the Western world before Hepatitis B became epidemic in the 1970s. HBV is also a blood-borne virus; therefore, shares the same transmission vectors and high-risk groups as HIV.

As for the legitimate forefront of the AIDS epidemic, now generally recognized as June 1981, a retrospective analysis of the New York State Cancer Registry identified two plausible AIDS-related cases of Kaposi's sarcoma diagnosed in two male residents of New York City in May and November of 1977, possibly pushing back the first dates of HIV infection to 1974 or so, in the opinion of the analyst.[454]

[In the opinion of the Author, the limited clinical and historical data presented on these two male NYC residents suggests that these 2 AIDS cases might be legitimate AIDS cases. However, what is missing in this period are any reports of PCP deaths. The prevalence of PCP and consequent deaths in following populations consistently exceeded those of KS. Patients who presented with KS tended to have slower disease progression onto death than patients who presented with PCP. The PCP deaths of this time period, if they exist, remain unrecorded.]

Other Types of Immunodeficiency

Outliers are valid in their own domain but, by definition, they belong to a separate population(s). It is likely that some outliers cited as African AIDS cases actually had some legitimate form of immunodeficiency but, most likely, of a different character and etiology than HIV infection.

Immunodeficiency was exceedingly rare prior to the advent of AIDS, but not unknown. Immunodeficiency can be caused by substances or medical conditions and/or induced by congenital defects. **Aplastic** ("failure to thrive") anemia is one such example. The "boy in the bubble" – a cultural icon in itself, subject of both dramatic and comical movies – was firmly implanted into public consciousness by children with immunodeficiencies due to aplastic and **neoplastic** (cancerous) conditions.

Due to recent advances in immunology and genetics, many aplastic conditions have been attributed to specific genetic defects and labeled as **primary immunodeficiency disorders** (**PID**). More than 150 specific, genetic defects contributing to primary immunodeficiency have been identified.[455, 456] The majority of PIDs are humoral disorders: the patients lack the ability to produce antibodies, making them susceptible to bacterial infections (humoral immunity). Yet a subgroup of PIDs, **combined variable immunodeficiency disorders** (**CVIDs**), consist of a heterogeneous group of disorders having both impaired cell-mediated immunity (T lymphocytes) and humoral immunity (antibodies). The prevalence of CVIDs is approximately one per 50,000 live births.[456]

Opportunistic pathogens that take advantage of the immune defect induced by CVID include *Candida*, *Cryptococcus neoformans*, *Pneumocystis carinii*, *Toxoplasma gondii*, and *Cryptosporidium*: the first four being AIDS-defining conditions, and the latter being a frequent enteric pathogen of AIDS patients.[456] CVID may present at any age, but the peak incidence seems to occur in the first and third decades of life.[457, 458] Median age of diagnosis was 23 years for males and 28 years for females in one case series ($n = 284$).[458]

Cancers within components of the immune system can also induce immunodeficiency. (Such cancers affect the bone marrow, the lymph glands, or the white blood cell populations.) For instance, *Pneumocystis carinii* pneumonia can also be a consequence of leukemia, lymphoma, or solid tumors.[459-461] Certain medical drug regimens can induce immunodeficiency. Furthermore, immunodeficiency can be transmitted by viruses other than HIV.

For example, infection by at least one other virus can lead to fatal *Pneumocystis carinii* pneumonia in an animal model. Two infant chimpanzees died from *Pneumocystis carinii* pneumonia and erythroleukemia (proliferative disorder of germinal red and white blood cell tissues) after drinking milk from a cow infected with **bovine leukemia virus** (BLV), a retrovirus (but a different Genus than HIV). Four other chimpanzee infants thrived unharmed.[462] Could rare cases of human immunodeficiency have occurred in the same way – people drinking un-pasteurized milk from cows infected with BLV: a method of viral transmission that might produce clusters?

In fact, there are statistical hints that exposure to the blood of ungulates carries an increased risk of immunological problems. Specifically, working in an abattoir may carry an increased risk of leukemia, and exposure to meat in a processing facility carries an increased risk of lymphomas. The presence of carcinogenic chemicals in the processing facility could also be a factor.[463-467] (These statistical analyses revealed an association between the risk of leukemia and lymphomas and exposure to the blood or meat of ungulates, but such basic statistical analyses are not able to prove definitive cause.)

Malnutrition is a recognized cause of immunodeficiency. Malnutrition impairs cell-mediated immunity to the extent that three diseases considered "AIDS-defining" diseases are also frequent opportunistic diseases secondary to malnutrition; namely, *Pneumocystis carinii* pneumonia (PCP), candidiasis, and disseminated *Herpes* viral infections. Protein-calorie malnutrition, malnourishment, iron deficiency, and a diet high in carbohydrates have been associated with oral candidiasis; one can have protein-calorie malnutrition and not look gaunt.[468-472]

The impairment of cell-mediated immunity induced by protein-calorie malnutrition renders people more susceptible to several other diseases and conditions commonly associated with HIV/AIDS among purported African patients; namely tuberculosis, respiratory tract infections, diarrhea; and a reduced immunological response to mitogens (allergens). Malnutrition-induced immunodeficiency is also correlated with hepatitis, certain bacterial infections (gram-negative), a variety of dermatoses, and morbidity and mortality consequent to measles and its corneal and pneumonic complications.[472-475]

The severity of invasive amoebiasis is increased in incidence and severity by malnutrition.[366] The symptoms of amoebiasis include severe weight loss, nausea, vomiting and diarrhea. The combination of diarrhea with weight loss is most common symptoms among

purported African AIDS cases (as described in chapter: *"African versus Classical AIDS"*).

> ## *"Unexplained recurrent infections and diarrhea maybe manifestation of poor nutritional status . . ."*[474]

In children, protein-calorie malnutrition reduced tonsil size and the immunological

response to chemicals and mitogens. Protein-calorie malnutrition also induced chronic atrophy

of the thymus. Also, there was a wasting of peripheral lymphoid tissue, and depletion of

paracortical cells and loss of germinal centers.[472]

The thymus is where most T lymphocytes mature. The peripheral lymphoid tissue

includes the lymph nodes (glands) and lymphatic tissues in many other organs. The paracortical

cells are lymphatic tissues in the brain. Germinal centers are the sites in lymph nodes where B

lymphocytes mature. B lymphocytes generate antibodies; thus, an impairment in B lymphocyte

maturation impairs humoral immunity; making the patient more susceptible to bacterial

infections as seen in children with protein-malnutrition.

In mice, malnourishment induced atrophy of lymphatic tissue ("not a new observation"),

impaired immune response, and increased severity of virus infection. One group of researchers

found that with chronic undernutrition the lymphocyte counts of severely malnourished mice

could not be returned to normal levels by adrenalectomy (removal of the adrenal gland; whose

hormones control lymphatic tissue structure and function). In mice that had been moderately

malnourished, removal of the adrenal gland reversed the drop in lymphocyte counts. Thus, the

physical effects of malnutrition can apparently damage organ components of the immune system

beyond repair, at least in animal models.[476]

WHY THE CONCEPT OF A 10-YEAR INCUBATION PERIOD?

The Author believes that the concept of the 10-year incubation period for HIV infection arose from a general misunderstanding. By 1985 or so, a successful prophylaxis regimen had been developed for PCP, curtailing the primary cause of rapid AIDS mortality, and other drugs were becoming successful in staving off immediate death from other opportunistic infections. After this limited success, there was some speculation in medical fields that, given the growing efficiency of prophylaxis, the survival of AIDS patients from the time of infection until the death be extend to 10 years. This speculation was echoed by the general media. Thus, patients' *survival time* might have increased to 10 years, but the incubation period of untreated HIV infection would not have changed necessarily in absence of an effective prophylactic antiretroviral regimen which became available years later. The two concepts – incubation and survival – are typically melded into one.

Also, one HIV came to be recognized as a lentivirus, the concept of HIV having long, silent incubation period was inculcated into the general and professional domains. The pathogen lentiviruses known prior to the emergence of HIV, those that infected ungulates, induced slowly progressive diseases. In humans, a minimal average incubation period of 1 – 2 years for HIV infection is a long, silent incubation period compared with the typical incubation period of days or weeks for typical prevalent bacterial and viral infections.

The replication of HIV may progress slowly over this period, but once the T4-cell count drops below an average threshold (200 T4-cells/μL), both symptom onset and mortality can occur abruptly. For example, in the United States, among the first 159 AIDS patients, 40% died from their initial PCP infection.[380] In Britain, among the first 130 AIDS cases, mortality from the first attack of *Pneumocystis carinii* pneumonia (PCP) was about 33%.[477]

Chapter 16

When Hype Equals Hypotheses

The basic theorem of the origin-of-AIDS-in-Africa hypothesis is that HIV was endemic in Africa for 30 or 40 years before ecological and sociological changes forced it out of the jungle. Once exposed to naïve, urban, highly susceptible populations, the disease spread geometrically (repeatedly doubling in number in a given time period) to infect tens or hundreds of millions of Africans (without drawing notice!) before reaching across the Atlantic to selectively infect gay men in New York City.

It is evident that the principal epicenter of the AIDS epidemic was New York City. In the old days, disease traveled by sailing ships. AIDS flew. AIDS traveled from NYC to other party centers: San Francisco, Los Angeles, Miami, Washington DC, Berlin, London, and Copenhagen, Sao Paulo, and Paris. The air-link from New York City and Paris was particularly important, as Paris was a key secondary epicenter. Paris was the epicenter of a global Francophone network, a network of urban centers in the world of French-speaking countries. From Paris, AIDS spread to Brussels, Belgium; Kinshasa, Zaire; Lusaka, Rwanda; and also Port-au-Prince, Haiti (or perhaps Paris and Brussels could be considered joint epicenters). Haiti was being cross-pollinated by both New York City and Paris. The characteristics of the risk groups in each of these global locales varied according to local culture, socioeconomic graphics, taboos, sexual practices, and the availability of medical care or access to clean hypodermic syringes.

There is no valid evidence that HIV was endemic in Africa before 1979. The early generation, antibody tests simply hold no validity. The first reasonable African clusters in Europe were identified 4 years after the first clusters were identified in NYC. The monkey viruses have no particular **homology** (a similarity often attributable to common ancestral origin) with HIV. (An appropriate comparison of homology is far beyond the scope of this document, but to compare the most consistently reported parameter, the *pol* region, the homology of HIV-1,

HIV-2, SIV_{AGM}, visna virus, and CAEV all range from approximately 50–60%.[101, 102, 116, 164, 166, 176, 478]) [*] The *pol* region encodes **reverse transcriptase** (formerly known as RNA-dependent, DNA polymerase), the enzyme essential to retroviruses, something not found in DNA-based organisms (nor in all RNA-based viruses).

In Africa, contrary to transmission patterns worldwide, the only alleged risk factor was heterosexual sex. However, AIDS in Africa was not demonstrably present in the rural poor. Rather, AIDS was documented as an urban disease, predominately afflicting the apparently heterosexual, highly sexually active, upper class – a group able to travel between countries at will. Many of the first Africans diagnosed with AIDS were residents of Europe, or had come to Europe seeking medical treatment for their condition.[57-59, 205, 226, 398, 479]

The African scenario makes far more sense when one assumes that a majority of the first male Africans with legitimate AIDS were secretly gay or bisexual. In the 1970s, gays "came out"– but only in the First World, not in the Third. The most plausible explanation is that these urban males contracted HIV in Belgium (or Paris, London, New York), and then infected their wives and/or lovers in Belgium and/or Zaire; some of whom, consequently, gave birth to HIV-infected children.[50, 57, 480, 481] If one accepts this premise, then AIDS in Africa would fall into alignment with the rest of the world.

Owing to the widely publicized, sexual practices of the New York City gay scene, and the gay adherents who flocked to NYC from across the world, both the medical and general media focused on sexual behavior, obscuring the issue of poverty and its association with sickness and early death (life expectancy at birth in Haiti is 30 years; 55 years for Zaire).[482-484]

From the Author's perspective, the true circumstances in these regions remain clouded. Haiti has receded from notice in the medical literature, and the data set that has emanated from Africa is a confused mixture of findings. Amongst this data set, it is difficult to discern

[*] the Author is guilty of a patent over-simplifcation here

legitimate AIDS cases from unqualified clinical and/or serological conditions that correlated with false seropositivity and/or the surveillance definition.

AFRICAN SUPERMEN

Other than heterosexual sex, the list of hypothetical HIV transmission methods included re-used hypodermic syringes, scarification rituals, voodoo blood rituals, monkey hunting and eating, monkey bites, sex with monkeys, sex with Haitians, promiscuity, and female circumcision – all speculation: nothing is documented in the medical or scientific literature. Another explanation is that HIV transmission reportedly occurs at a higher rate among uncircumcised men. Widespread circumcision for African men is now being recommended, based on insubstantial statistical data.

The HIV transmission rates among African heterosexuals are absurdly high compared with those of American heterosexual couples (as described previously). Prior to the advent of AIDS, NYC gay males were surveyed as averaging 20 sex partners in a 6-month period, or approximately 1000 sexual partners in a life-time.[485-487] (Per the New York Times, most men among the first Kaposi's sarcoma cluster typically had as many as 10 sexual encounters a night for up to four times per week.)[488] Given the estimated HIV transmission rate of heterosexual sex (risk per exposure of 0.0005% for insertive vaginal intercourse; 0.001% for receptive vaginal intercourse; and 0.005% for receptive anal intercourse); in theory, African heterosexual males would have to exceed the promiscuity of NYC gay males by an improbable factor of 5 to 20! [28, 489, 490]

The reported differences in clinical presentation between African AIDS and classical AIDS may arise from exposure to differing sets of ambient pathogens. For example, one might assume that African populations were exposed to enteric parasites in contaminated drinking water (an assumption made from afar). Or, this clinical profile might be a reflection of the surveillance definition itself. Likewise, it could be that other concomitant diseases and/or medical conditions might be overlaying this population, but mistakenly diagnosed as AIDS. One concerned scientist criticized WHO for inflating AIDS statistics: "The danger of the AIDS epidemic is dwarfed by 3.5 million deaths from tuberculosis and 16.8 million deaths from malaria since the beginning of the AIDS epidemic. The frightening scenario looms that widespread, but curable, diseases are wrongly classified as AIDS-related complex, thereby foregoing appropriate treatment."[288]

In a letter to the *British Medical Journal*, another author described "the self-fulfilling prophecy of increasing deaths in Africa when incoming funds mainly go to AIDS prevention instead of combating treatable conditions that look like AIDS. "In Uganda, $57,000 a year is used in the control and treatment of malaria, compared with $6m for controlling AIDS."[491]

Undoubtedly, AIDS does exist in Africa. However, authentic African cases need to be differentiated from illogical statistical projections, just as they need to be differentiated from those patients "diagnosed" with AIDS but whose characteristics differ from classical AIDS in disease manifestation, risk-group status, and purported efficacy of transmission.

HOW DO THESE MISCONCEPTIONS HAPPEN?

The term "counting coup" is an antiquated American term denoting actions that grant one prestige. Etymologically, the term comes from the Native Americans (a.k.a. American Indians). "Coup" was gained by acts of bravery during battle, such as striking an enemy in battle with only one's hand or coup stick and escaping unharmed; or, being the first to touch a dead enemy with one's coup stick. The later recounting of these acts of bravery (or actions designed to gain one prestige) came to be known as "counting coup" in the American English vernacular.

Scientists count coup by finding facts substantiating popular and seemingly valid scientific theories, although through the compartmentalization of science via specialization, the interwoven supportive theories and/or findings gain a generalized sense of credibility beyond the limited findings within each individual domain.

Chapter 17

And then came HIV-2

The story of HIV-2 distinguishes HIV-1 from the primate viruses. The zoonotic reservoir of HIV-1 was purportedly chimpanzees (*Pan troglodytes troglodytes*). The zoonotic reservoir of HIV-2 was purportedly sooty mangabey monkeys (*Cercocebus atys*). HIV-2 seems to be endemic to Western Africa, which is also the habitat of the sooty mangabey; which are hunted and kept as household pets in rural areas. Epidemiologically, HIV-2 is an urban disease and linked with Portugal. Initially, HIV-2 was reported both as non-pathogenic *and* causing a disease similar to AIDS. Over time it became clear that HIV-2 is not as pathogenic as HIV-1.

In terms of homology, HIV-2 has portions that are 75% homologous with SIV_{SM} from sooty mangabeys from West African and also 75% homology with $STLV-III_{MAC}$, the laboratory contaminant which was the virus taken from 4 sick Rhesus macaque monkeys living in Southborough, Massachusetts. $STLV-III_{MAC}$ is likely a descendant of viruses brought over from Africa years ago along with the imported primate populations. But HIV-2 is only 50% homologous with HIV-1. Many researchers define phylogenic relationship of HIV-1 and HIV-2 as "close." A few define it as "distantly related."

Compared with HIV-1, the zoonotic theory for the emergence of HIV-2 has greater credibility. The purported zoonotic reservoir (sooty mangabeys) resides in the same geographic region as the apparent disease epicenter, although HIV-2 disease was first noticed among urban areas of West Africa. Several of the first patients diagnosed with HIV-2 infection had also resided extensively in Europe.

A further clarification of HIV-2, related pathological conditions, epidemiology, and homology with HIV-1 will be discussed in the forthcoming book.

Errata & Updates

This work was a labor of love that taxed everyone involved beyond all reasonable bounds. Undoubtedly, it contains grammatical, factual, citational, and conceptual errors, so please notify us when you find them (www.healthalert.net/Contact.html). Errata and updates will be posted updated periodically at: www.healthalert.net/TFTF_errata&updates.pdf.

Glossary

AIDS prodrome

An indeterminate period generally including lymphadenopathy, fever, weight loss, vague malaise: prior to the clinical presentation of opportunistic infections, generally accompanied by cutaneous anergy and disturbed lymphocyte profile

Combined variable immunodeficiency disorders (CVIDs)

A subgroup of primary immunodeficiency disorders (PIDs), CVIDs consist of a heterogeneous group of disorders having both impaired cell-mediated immunity (T-lymphocytes) and humoral immunity (antibodies). Due to impairments of the cell-mediated immunity, patients with CVID are susceptible to several diseases that would otherwise define AIDS.

Endometritis

Inflammation of the endometrium, the mucous membrane lining the uterus

Factor VIII

A protein that is an essential clotting factor of blood plasma that is absent or inactive in hemophilia

Immunofluorescence assay

In this assay, antibodies marked with a fluorescent dye; the antibodies bind to viral antigens, and the marked antibodies are viewed via light microscopy

Immunoprecipitation assay

An immunoprecipitation assay is the process of precipitating ("crystallizing") a protein antigen out of a liquid using an antibody that specifically binds to that particular protein

Lymphatic system

The lymphatic system, the human body's second fluid system, is part of the body's immune system, and contains the clear fluid **lymph**. The lymphatic system drains fluid out of the body's tissues. Lymph nodes (glands) are filtering stations for lymphatic fluid, extracting bacteria, foreign substances, and dead white blood cells from the fluid. In each lymph node, a variety of compartments contain T-cells, **B-cells** (which secrete antibodies), and **macrophages** (which engulf infected cells marked with foreign antigens). T-cells can migrate back and forth between the blood system and the lymphatic system, perhaps introducing HIV into the **thymus**. Thus all the factors needed for an immune response are brought together in lymph nodes. With all this activity, lymph nodes often swell during infection. This nodal swelling is why doctors feel the lymph nodes around the neck, under the armpits, and in the groin during physical checkups. A healthy lymphatic system is required for an effective immune system. HIV can also infect macrophages. Macrophages can cross the blood-brain barrier and introduce HIV into the central nervous system

Notifiable disease

Any disease required by law to be reported to government authorities

Phylogeny

The evolutionary history of a particular taxonomic group; taxonomy relates to the nomenclature, description, and classification of organisms, and the study of their relationship

Positive Predictive Value

The probability that a patient with a positive test result really does have the condition for which the test was conducted. Also related to the negative predictive value, the probability that a patient with a negative test result really is free of the condition for which the test was conducted.

Retrovirus

A retrovirus is an RNA virus encodes genetic information in an RNA rather than DNA; however, they utilize a DNA intermediate during replication. Retroviruses use the enzyme reverse transcriptase to reverse-transcribe ("reverse-write) DNA from RNA. The DNA is then incorporated into the host's genome. This process essentially "hi-jacks" the host cell: turning the host cell into a virus-making factory. In DNA-based organisms, such as humans, enzymes transcribe ("write") RNA from DNA, i.e., the "normal" process.

Restriction patterns

The number and sizes of the DNA fragments produced when a particular DNA molecule is cut with a particular enzyme

Syncytia

Syncytia are amorphous single-cell masses of cytoplasm containing multiple nuclei: cellular abnormalities formed by the fusion of multiple cells

Taxonomy

Orderly classification of plants and animals according to their presumed natural relationships

References

1. Siegal FP, Lopez C, Hammer GS, et al. Severe acquired immunodeficiency in male homosexuals, manifested by chronic perianal ulcerative herpes simplex lesions. *N Engl J Med.* Dec 10 1981;305(24):1439-1444.

2. Gottlieb MS, Schroff R, Schanker HM, et al. Pneumocystis carinii pneumonia and mucosal candidiasis in previously healthy homosexual men: evidence of a new acquired cellular immunodeficiency. *N Engl J Med.* Dec 10 1981;305(24):1425-1431.

3. CDC. Pneumocystis Pneumonia - Los Angeles. *MMWR Morb Mortal Wkly Rep.* June 5 1981;30(21):250-252.

4. Masur H, Michelis MA, Greene JB, et al. An outbreak of community-acquired Pneumocystis carinii pneumonia: initial manifestation of cellular immune dysfunction. *N Engl J Med.* Dec 10 1981;305(24):1431-1438.

5. CDC. Kaposi's sarcoma and Pneumocystis pneumonia among homosexual men--New York City and California. *MMWR Morb Mortal Wkly Rep.* Jul 3 1981;30(25):305-308.

6. CDC. Follow-up on Kaposi's sarcoma and Pneumocystis pneumonia. *MMWR Morb Mortal Wkly Rep.* Aug 28 1981;30(33):409-410.

7. CDC. Update on acquired immune deficiency syndrome (AIDS)--United States. *MMWR Morb Mortal Wkly Rep.* Sep 24 1982;31(37):507-508, 513-514.

8. CDC. Kaposi's Sarcoma (KS), Pneumocystis Carinii Pneumonia (PCP), and Other Opportunistic Infections (OI): Cases Reported to CDC as sf July 8, 1982. *HIV Surveillance Report* [http://www.cdc.gov/hiv/topics/surveillance/resources/reports/pdf/surveillance82.pdf.

9. CDC. Acquired Immunodeficiency Syndrome (AIDS) Weekly Surveillance Report - United States, December 31, 1984. *HIV Surveillance Report* [http://www.cdc.gov/hiv/topics/surveillance/resources/reports/pdf/surveillance84.pdf.

10. CDC. Acquired Immunodeficiency Syndrome (AIDS) Weekly Surveillance Report - United States, December 30, 1985. *HIV Surveillance Report* [http://www.cdc.gov/hiv/topics/surveillance/resources/reports/pdf/surveillance85.pdf.

11. Jaffe HW, Bregman DJ, Selik RM. Acquired immune deficiency syndrome in the United States: the first 1,000 cases. *J Infect Dis.* Aug 1983;148(2):339-345.

12. Altman KA. Rare Cancer seen in 41 Homosexuals. *New York Times.* July 3, 1981, 1981.

13. Medicine: Opportunistic Diseases. *Time.* New York, NY; 1981.

14. CDC. Acquired Immunodeficiency Syndrome (AIDS) Weekly Surveillance Report - United States, December 22, 1983. *HIV Surveillance Report* [http://www.cdc.gov/hiv/topics/surveillance/resources/reports/pdf/surveillance83.pdf.

15. CDC. Acquired immunodeficiency syndrome (AIDS) update--United States. *MMWR Morb Mortal Wkly Rep.* Jun 24 1983;32(24):309-311.

16. CDC. Update: acquired immunodeficiency syndrome (AIDS)--United States. *MMWR Morb Mortal Wkly Rep.* Sep 9 1983;32(35):465-467.

17. CDC. Update: acquired immunodeficiency syndrome (AIDS) - United States. *MMWR Morb Mortal Wkly Rep.* Jan 6 1984;32(52):688-691.

18. CDC. Update: acquired immunodeficiency syndrome (AIDS)--United States. *MMWR Morb Mortal Wkly Rep.* Jun 22 1984;33(24):337-339.

19. Klein RS, Harris CA, Small CB, Moll B, Lesser M, Friedland GH. Oral candidiasis in high-risk patients as the initial manifestation of the acquired immunodeficiency syndrome. *N Engl J Med.* Aug 9 1984;311(6):354-358.

20. Reichert CM, O'Leary TJ, Levens DL, Simrell CR, Macher AM. Autopsy pathology in the acquired immune deficiency syndrome. *Am J Pathol.* Sep 1983;112(3):357-382.

21. Welch K, Finkbeiner W, Alpers CE, et al. Autopsy findings in the acquired immune deficiency syndrome. *JAMA.* Sep 7 1984;252(9):1152-1159.
22. Holmberg K, Meyer RD. Fungal infections in patients with AIDS and AIDS-related complex. *Scand J Infect Dis.* 1986;18(3):179-192.
23. Barr CE, Torosian JP. Oral manifestations in patients with AIDS or AIDS-related complex. *Lancet.* Aug 2 1986;2(8501):288.
24. Moore RD, Chaisson RE. Natural history of opportunistic disease in an HIV-infected urban clinical cohort. *Ann Intern Med.* Apr 1 1996;124(7):633-642.
25. Haverkos HW, Drotman DP. Prevalence of Kaposi's sarcoma among patients with AIDS. *N Engl J Med.* Jun 6 1985;312(23):1518.
26. Safai B, Johnson KG, Myskowski PL, et al. The natural history of Kaposi's sarcoma in the acquired immunodeficiency syndrome. *Ann Intern Med.* Nov 1985;103(5):744-750.
27. Beral V, Bull D, Darby S, et al. Risk of Kaposi's sarcoma and sexual practices associated with faecal contact in homosexual or bisexual men with AIDS. *Lancet.* Mar 14 1992;339(8794):632-635.
28. Smith DK, Grohskopf LA, Black RJ, et al. Antiretroviral postexposure prophylaxis after sexual, injection-drug use, or other nonoccupational exposure to HIV in the United States: recommendations from the U.S. Department of Health and Human Services. *MMWR Recomm Rep.* Jan 21 2005;54(RR-2):1-20.
29. CDC. Table 19 - Reported AIDS cases, by transmission category and sex, 2007 and cumulative—United States and dependent areas. *HIV/AIDS Surveillance Report* [http://www.cdc.gov/hiv/surveillance/resources/reports/2007report/table19.htm. Accessed October 24, 2010.
30. CDC. U.S. HIV and AIDS cases reported through December 1993. *HIV/AID Surveillance Report.* 1993;5(5).
31. CDC. *AIDS Weekly Surveillance Report - September 21, 1987* September 21 1987.
32. Thomsen HK, Jacobsen M, Malchow-Moller A. Kaposi sarcoma among homosexual men in Europe. *Lancet.* Sep 26 1981;2(8248):688.
33. Gerstoft J, Malchow-Moller A, Bygbjerg I, et al. Severe acquired immunodeficiency in European homosexual men. *Br Med J (Clin Res Ed).* Jul 3 1982;285(6334):17-19.
34. Melbye M, Biggar RJ, Ebbesen P, et al. Seroepidemiology of HTLV-III antibody in Danish homosexual men: prevalence, transmission, and disease outcome. *Br Med J (Clin Res Ed).* Sep 8 1984;289(6445):573-575.
35. Biggar RJ, Melbye M, Ebbesen P, et al. Low T-lymphocyte ratios in homosexual men. Epidemiologic evidence for a transmissible agent. *Jama.* Mar 16 1984;251(11):1441-1446.
36. Gerstoft J, Nielsen JO, Dickmeiss E, Ronne T, Platz P, Mathiesen L. The acquired immunodeficiency syndrome (AIDS) in Denmark. A report from the Copenhagen study group of AIDS on the first 20 Danish patients. *Acta Med Scand.* 1985;217(2):213-224.
37. L'Age-Stehr J, Kunze R, Koch MA. AIDS in West Germany. *Lancet.* Dec 10 1983;2(8363):1370-1371.
38. du Bois RM, Branthwaite MA, Mikhail JR, Batten JC. Primary Pneumocystis carinii and cytomegalovirus infections. *Lancet.* Dec 12 1981;2(8259):1339.
39. Surveillance of the acquired immune deficiency syndrome in the United Kingdom, January 1982-July 1983. *Br Med J (Clin Res Ed).* Aug 6 1983;287(6389):407-408.
40. Weller I, Crawford DH, Iliescu V, et al. Homosexual men in London: lymphadenopathy, immune status, and Epstein-Barr virus infection. *Ann N Y Acad Sci.* 1984;437:238-253.
41. CDC. Update: acquired immunodeficiency syndrome (AIDS)--United States. *MMWR Morb Mortal Wkly Rep.* Aug 5 1983;32(30):389-391.

42. Winterton SJ, Warren JB, Barnes PJ. UK case of AIDS in an intravenous drug abuser. *Lancet.* May 25 1985;1(8439):1223.

43. Davison AG, du Bois RM. Persistent generalised lymphadenopathy in a female in the United Kingdom. *Lancet.* May 25 1985;1(8439):1223.

44. CDC. Update: acquired immunodeficiency syndrome--United States. *MMWR Morb Mortal Wkly Rep.* May 10 1985;34(18):245-248.

45. Rozenbaum W, Coulaud JP, Saimot AG, Klatzmann D, Mayaud C, Carette MF. Multiple opportunistic infection in a male homosexual in France. *Lancet.* Mar 6 1982;1(8271):572-573.

46. Brunet JB, Bouvet E, Leibowitch J, et al. Acquired immunodeficiency syndrome in France. *Lancet.* Mar 26 1983;1(8326 Pt 1):700-701.

47. CDC. Current Trends Prevention of Acquired Immune Deficiency Syndrome (AIDS): Report of Inter-Agency Recommendations *MMWR Morb Mortal Wkly Rep.* March 04, 1983 1983;32(8):101-103.

48. De Cock KM. AIDS: an old disease from Africa? *Br Med J (Clin Res Ed).* Aug 4 1984;289(6440):306-308.

49. Bygbjerg IC. AIDS in a Danish surgeon (Zaire, 1976). *Lancet.* Apr 23 1983;1(8330):925.

50. Brunet JB, Bouvet E, Massari V. Epidemiological aspects of acquired immune deficiency syndrome in France. *Ann N Y Acad Sci.* 1984;437:334-339.

51. Mathez D, Leibovitch J, Sultan Y, Maisonneuve P. LAV/HTLV-III seroconversion and disease in hemophiliacs treated in France. *N Engl J Med.* Jan 9 1986;314(2):118-119.

52. Melbye M, Froebel KS, Madhok R, et al. HTLV-III seropositivity in European haemophiliacs exposed to Factor VIII concentrate imported from the USA. *Lancet.* Dec 22 1984;2(8417-8418):1444-1446.

53. Immunologic and virologic status of multitransfused patients: role of type and origin of blood products. By the AIDS-Hemophilia French Study Group. *Blood.* Oct 1985;66(4):896-901.

54. Blood transfusion, haemophilia, and AIDS. *Lancet.* Dec 22 1984;2(8417-8418):1433-1435.

55. Ras GJ, Simson IW, Anderson R, Prozesky OW, Hamersma T. Acquired immunodeficiency syndrome. A report of 2 South African cases. *S Afr Med J.* Jul 23 1983;64(4):140-142.

56. Edwards D, Harper PG, Pain AK, Welch J, Barbatis C, Mallinson C. Kaposi's sarcoma associated with AIDS in a woman from Uganda. *Lancet.* Mar 17 1984;1(8377):631-632.

57. Van de Perre P, Rouvroy D, Lepage P, et al. Acquired immunodeficiency syndrome in Rwanda. *Lancet.* Jul 14 1984;2(8394):62-65.

58. Piot P, Quinn TC, Taelman H, et al. Acquired immunodeficiency syndrome in a heterosexual population in Zaire. *Lancet.* Jul 14 1984;2(8394):65-69.

59. Clumeck N, Sonnet J, Taelman H, et al. Acquired immunodeficiency syndrome in African patients. *N Engl J Med.* Feb 23 1984;310(8):492-497.

60. Lyons SF, Schoub BD, McGillivray GM, Sher R, Dos Santos L. Lack of evidence of HTLV-III endemicity in southern Africa. *N Engl J Med.* May 9 1985;312(19):1257-1258.

61. Schoub BD, Lyons SF, McGillivray GM, Smith AN, Johnson S, Fisher EL. Absence of HIV infection in prostitutes and women attending sexually-transmitted disease clinics in South Africa. *Trans R Soc Trop Med Hyg.* 1987;81(5):874-875.

62. CDC. Cases of specified notifiable disease,United States, weeks ending April 11, 1987 and April 5, 1986 (14th Week). *MMWR Morb Mortal Wkly Rep.* April 17, 1987 1987;36(14):215.

63. CDC. Acquired Immunodeficiency Syndrome (AIDS) Weekly Surveillance Report - United States, December 29, 1986. http://www.cdc.gov/hiv/topics/surveillance/resources/reports/pdf/surveillance86.pdf.

64. Cohn RJ, MacPhail AP, Hartman E, Schwyzer R, Sher R. Transfusion-related human immunodeficiency virus in patients with haemophilia in Johannesburg. *S Afr Med J.* Dec 1 1990;78(11):653-656.

65. CDC. Opportunistic infections and Kaposi's sarcoma among Haitians in the United States. *MMWR Morb Mortal Wkly Rep.* Jul 9 1982;31(26):353-354, 360-351.

66. Vieira J, Frank E, Spira TJ, Landesman SH. Acquired immune deficiency in Haitians: opportunistic infections in previously healthy Haitian immigrants. *N Engl J Med.* Jan 20 1983;308(3):125-129.

67. Pitchenik AE, Fischl MA, Dickinson GM, et al. Opportunistic infections and Kaposi's sarcoma among Haitians: evidence of a new acquired immunodeficiency state. *Ann Intern Med.* Mar 1983;98(3):277-284.

68. Johnson WD, Jr., Pape JW. AIDS in Haiti. *Immunol Ser.* 1989;44:65-78.

69. Malebranche R, Arnoux E, Guerin JM, et al. Acquired immunodeficiency syndrome with severe gastrointestinal manifestations in Haiti. *Lancet.* Oct 15 1983;2(8355):873-878.

70. Fischl MA, Dickinson GM. An acquired immunodeficiency syndrome among Haitians: an update. *Ann N Y Acad Sci.* 1984;437:325-333.

71. Pape JW, Liautaud B, Thomas F, et al. Characteristics of the acquired immunodeficiency syndrome (AIDS) in Haiti. *N Engl J Med.* Oct 20 1983;309(16):945-950.

72. Pape JW, Johnson W. Epidemiology of AIDS in the Caribbean. In: Piot P, Mann J, eds. *Balliere's Clinical Tropical. Medicine and Communicable Diseases.* Vol 3; 1988:31-42.

73. Wenk RE, Russo F. Group-specific component (Gc) alleles among Haitian immigrants. *Am J Clin Pathol.* Sep 1988;90(3):371-372.

74. Moses P, Moses J. Haiti and the Acquired Immunodeficiency Syndrome. *Ann Intern Med.* October 1 1983;99(4):565.

75. Greenfield WR. Night of the living dead II: slow virus encephalopathies and AIDS: do necromantic zombiists transmit HTLV-III/LAV during voodooistic rituals? *JAMA.* Oct 24-31 1986;256(16):2199-2200.

76. Leonidas JR, Hyppolite N. Haiti and the acquired immunodeficiency syndrome. *Ann Intern Med.* Jun 1983;98(6):1020-1021.

77. Henig M. AIDS - A New Disease's Deadly Odyssey. *New York Times.* February 6, 1983, 1983.

78. Pape JW, Liautaud B, Thomas F, et al. The acquired immunodeficiency syndrome in Haiti. *Ann Intern Med.* Nov 1985;103(5):674-678.

79. Deschamps MD. AIDS in the Caribbean. *Arch AIDS Res.* 1988;2(1):51-56.

80. Barry M, Stansfield SK, Bia FJ. Haiti and the Hopital Albert Schweitzer. *Ann Intern Med.* Jun 1983;98(6):1018-1020.

81. Guerin JM, Malebranche R, Elie R, et al. Acquired immune deficiency syndrome: specific aspects of the disease in Haiti. *Ann N Y Acad Sci.* 1984;437:254-263.

82. Greco RS. Haiti and the stigma of AIDS. *Lancet.* Aug 27 1983;2(8348):515-516.

83. Altema R, Bright L. Only homosexual Haitians, not all Haitians. *Ann Intern Med.* Dec 1983;99(6):877-878.

84. Lange WR, Jaffe JH. AIDS in Haiti. *N Engl J Med.* May 28 1987;316(22):1409-1410.

85. Liautaud B, Pape J, Pamphile M. Le SIDA dans les caraibes *Médecine et Maladies Infectieuses.* December 1988;18(Suppl 5):687-697.

86. Farmer P. *AIDS and accusation: Haiti and the geography of blame.* Berkeley and Los Angeles, California: University of California Press; 1992.

87. Molinert HT, Galban Garcia E, Rodriguez Cruz R. Prevalence of infection with human immunodeficiency virus in Cuba. *Bull Pan Am Health Organ.* 1989;23(1-2):62-67.

88. Brunet JB, Ancelle RA. The international occurrence of the acquired immunodeficiency syndrome. *Ann Intern Med.* Nov 1985;103(5):670-674.

89. Rich V. AIDS: Poland's minister for prophylaxis. *Nature.* September 12, 1985 1985;317(6033):100.

90. Savitskii SN, Irova TI, Vostrikova EP, Denisov A, Red'kin AP. [Renal lesions in patients infected with human immunodeficiency virus]. *Ter Arkh.* 1988;60(6):127-129.

91. Pokrovskii VI, Pokrovskii VV, Potekaev NS, Karetkina E, Astaf'eva NV. [The first case of acquired immunodeficiency syndrome in an USSR citizen]. *Ter Arkh.* 1988;60(7):10-14.

92. Pokrovskii VV, Eramova I. [The penetration of HIV into the population of Moscow homosexuals and its spread]. *Zh Mikrobiol Epidemiol Immunobiol.* May 1990(5):18-22.

93. CDC. U.S. AIDS cases reported through December 1989. *HIV/AIDS Surveillance* [Year End Edition:http://www.cdc.gov/hiv/topics/surveillance/resources/reports/pdf/surveillance89.pdf.

94. CDC. Diagnoses of HIV infection and AIDS in the United States and Dependent Areas, 2008. Vol 20. Atlanta GA, USA: Centers for Disease Control and Prevention, Department of Health and Human Services; 2010.

95. Barre-Sinoussi F, Chermann JC, Rey F, et al. Isolation of a T-lymphotropic retrovirus from a patient at risk for acquired immune deficiency syndrome (AIDS). *Science.* May 20 1983;220(4599):868-871.

96. Gallo RC, Sarin PS, Gelmann EP, et al. Isolation of human T-cell leukemia virus in acquired immune deficiency syndrome (AIDS). *Science.* May 20 1983;220(4599):865-867.

97. Wong-Staal F, Hahn BH, Shaw GM, et al. Molecular characterization of human T-lymphotropic leukemia virus type III associated with the acquired immunodeficiency syndrome. *Princess Takamatsu Symp.* 1984;15:291-300.

98. Shaw GM, Hahn BH, Arya SK, Groopman JE, Gallo RC, Wong-Staal F. Molecular characterization of human T-cell leukemia (lymphotropic) virus type III in the acquired immune deficiency syndrome. *Science.* Dec 7 1984;226(4679):1165-1171.

99. Lee TH, Coligan JE, Essex M. Human T-cell leukemia virus specific antigens. *Princess Takamatsu Symp.* 1984;15:197-203.

100. Grantham P, Perrin P. AIDS virus and HTLV-I differ in codon choices. *Nature.* Feb 27-Mar 5 1986;319(6056):727-728.

101. Chiu IM, Yaniv A, Dahlberg JE, et al. Nucleotide sequence evidence for relationship of AIDS retrovirus to lentiviruses. *Nature.* Sep 26-Oct 2 1985;317(6035):366-368.

102. Gonda MA, Wong-Staal F, Gallo RC, Clements JE, Narayan O, Gilden RV. Sequence homology and morphologic similarity of HTLV-III and visna virus, a pathogenic lentivirus. *Science.* Jan 11 1985;227(4683):173-177.

103. Gonda MA, Wong-Staal F, Gallo RC, Clements JE, Gilden RV. Heteroduplex mapping in the molecular analysis of the human T-cell leukemia (lymphotropic) viruses. *Cancer Res.* Sep 1985;45(9 Suppl):4553s-4558s.

104. Chang SY, Bowman BH, Weiss JB, Garcia RE, White TJ. The origin of HIV-1 isolate HTLV-IIIB. *Nature.* Jun 3 1993;363(6428):466-469.

105. Vahlne A. A historical reflection on the discovery of human retroviruses. *Retrovirology.* 2009;6:40.

106. Kanki PJ, Alroy J, Essex M. Isolation of T-lymphotropic retrovirus related to HTLV-III/LAV from wild-caught African green monkeys. *Science.* Nov 22 1985;230(4728):951-954.

107. Kanki PJ, Barin F, M'Boup S, et al. New human T-lymphotropic retrovirus related to simian T-lymphotropic virus type III (STLV-IIIAGM). *Science.* Apr 11 1986;232(4747):238-243.

108. Boffey P. U.S. and French teams report AIDS virus finds. *New York Times.* March 27, 1986, 1986;U.S.

109. Altman KA. Linking AIDS to Africa provokes bitter debate. *New York Times.* November 21, 1985, 1985;U.S.

110. Daniel MD, Letvin NL, King NW, et al. Isolation of T-cell tropic HTLV-III-like retrovirus from macaques. *Science.* Jun 7 1985;228(4704):1201-1204.

111. Mulder C. Virology. A case of mistaken non-identity. *Nature.* Feb 18 1988;331(6157):562-563.

112. Desrosiers RC. Origin of the human AIDS virus. *Nature.* Feb 27-Mar 5 1986;319(6056):728.

113. Kestler HW, 3rd, Li Y, Naidu YM, et al. Comparison of simian immunodeficiency virus isolates. *Nature.* Feb 18 1988;331(6157):619-622.

114. Mulder C. Human AIDS virus not from monkeys. *Nature.* Jun 2 1988;333(6172):396.

115. Hirsch V, Riedel N, Kornfeld H, Kanki PJ, Essex M, Mullins JI. Cross-reactivity to human T-lymphotropic virus type III/lymphadenopathy-associated virus and molecular cloning of simian T-cell lymphotropic virus type III from African green monkeys. *Proc Natl Acad Sci U S A.* Dec 1986;83(24):9754-9758.

116. Franchini G, Gurgo C, Guo HG, et al. Sequence of simian immunodeficiency virus and its relationship to the human immunodeficiency viruses. *Nature.* Aug 6-12 1987;328(6130):539-543.

117. Hirsch V, Riedel N, Mullins JI. The genome organization of STLV-3 is similar to that of the AIDS virus except for a truncated transmembrane protein. *Cell.* May 8 1987;49(3):307-319.

118. Newmark P. Variations of AIDS virus relatives. *Nature.* Apr 9-15 1987;326(6113):548.

119. Desrosiers RC, Daniel MD, Letvin NL, King NW, Hunt RD. Origins of HTLV-4. *Nature.* May 14-20 1987;327(6118):107.

120. Chakrabarti L, Guyader M, Alizon M, et al. Sequence of simian immunodeficiency virus from macaque and its relationship to other human and simian retroviruses. *Nature.* Aug 6-12 1987;328(6130):543-547.

121. Essex M, Kanki P. Reply to "Comparison of simian immunodefiency virus isolates". *Nature.* February 18, 1988 1988;331(6157):621-622.

122. Essex M, Kanki PJ. The origins of the AIDS virus. *Sci Am.* Oct 1988;259(4):64-71.

123. Ohta Y, Masuda T, Tsujimoto H, et al. Isolation of simian immunodeficiency virus from African green monkeys and seroepidemiologic survey of the virus in various non-human primates. *Int J Cancer.* Jan 15 1988;41(1):115-122.

124. Fultz PN, McClure HM, Anderson DC, Swenson RB, Anand R, Srinivasan A. Isolation of a T-lymphotropic retrovirus from naturally infected sooty mangabey monkeys (Cercocebus atys). *Proc Natl Acad Sci U S A.* Jul 1986;83(14):5286-5290.

125. Peeters M, Courgnaud V. *Overview of Primate Lentiviruses and Their Evolution in Non-human Primates in Africa*: Los Alamos National Laboratory, Los Alamos, New Mexico; 2002. LA-UR 03-3564.

126. Daniel MD, King NW, Letvin NL, Hunt RD, Sehgal PK, Desrosiers RC. A new type D retrovirus isolated from macaques with an immunodeficiency syndrome. *Science.* Feb 10 1984;223(4636):602-605.

127. Bryant ML, Gardner MB, Marx PA, et al. Immunodeficiency in rhesus monkeys associated with the original Mason-Pfizer monkey virus. *J Natl Cancer Inst.* Oct 1986;77(4):957-965.

128. Fine DL, Landon JC, Pienta RJ, et al. Responses of infant rhesus monkeys to inoculation with Mason-Pfizer monkey virus materials. *J Natl Cancer Inst.* Mar 1975;54(3):651-658.

129. Nowinski RC, Edynak E, Sarkar NH. Serological and structural properties of Mason-Pfizer monkey virus isolated from the mammary tumor of a Rhesus monkey. *Proc Natl Acad Sci U S A.* Jul 1971;68(7):1608-1612.

130. Fine DL. Mason-Pfizer monkey virus and simian AIDS. *Lancet.* Feb 11 1984;1(8372):335.

131. Benveniste RE, Morton WR, Clark EA, et al. Inoculation of baboons and macaques with simian immunodeficiency virus/Mne, a primate lentivirus closely related to human immunodeficiency virus type 2. *J Virol.* Jun 1988;62(6):2091-2101.

132. Lewis MG, Zack PM, Elkins WR, Jahrling PB. Infection of rhesus and cynomolgus macaques with a rapidly fatal SIV (SIVSMM/PBj) isolate from sooty mangabeys. *AIDS Res Hum Retroviruses.* Sep 1992;8(9):1631-1639.

133. Murphey-Corb M, Martin LN, Rangan SR, et al. Isolation of an HTLV-III-related retrovirus from macaques with simian AIDS and its possible origin in asymptomatic mangabeys. *Nature.* May 22-28 1986;321(6068):435-437.

134. Hirsch VM, Dapolito G, Johnson PR, et al. Induction of AIDS by simian immunodeficiency virus from an African green monkey: species-specific variation in pathogenicity correlates with the extent of in vivo replication. *J Virol.* Feb 1995;69(2):955-967.

135. Novembre FJ, De Rosayro J, O'Neil SP, Anderson DC, Klumpp SA, McClure HM. Isolation and characterization of a neuropathogenic simian immunodeficiency virus derived from a sooty mangabey. *J Virol.* Nov 1998;72(11):8841-8851.

136. Gao F, Bailes E, Robertson DL, et al. Origin of HIV-1 in the chimpanzee Pan troglodytes troglodytes. *Nature.* Feb 4 1999;397(6718):436-441.

137. Coffin J, Haase A, Levy JA, et al. Human immunodeficiency viruses. *Science.* May 9 1986;232(4751):697.

138. Coffin J, Haase A, Levy JA, et al. What to call the AIDS virus? *Nature.* May 1-7 1986;321(6065):10.

139. Shah K, Nathanson N. Human exposure to SV40: review and comment. *Am J Epidemiol.* Jan 1976;103(1):1-12.

140. O'Hearn EM. Profiles of Pioneer Women Scientists. Washington, DC: Acropolis Books Let; 1985:150-159, 160-169.

141. Rossiter MW. Women Scientists in America Before Affirmative Action: 1940-1972. *Women Scientists in America Before Affirmative Action: 1940-1972.* Baltimore and London: The John Hopkins University Press; 1995:288-289.

142. Butel JS, Lednicky JA. RESPONSE: re: cell and molecular biology of simian virus 40: implications for human infections and disease. *J Natl Cancer Inst.* Jul 7 1999;91(13):1166A-1167.

143. Butel JS, Lednicky JA. Cell and molecular biology of simian virus 40: implications for human infections and disease. *J Natl Cancer Inst.* Jan 20 1999;91(2):119-134.

144. Committee on Government Reform SoHRaW, 108th Congress, 1st Session. SV-40 Virus: Has Tainted Polio Vaccine Caused an Increase in Cancer, Hearing, September 10, 2003 Committee on Government Reform, Subcommittee on Human Rights and Wellness; 2004.

145. Kanki PJ, McLane MF, King NW, Jr., et al. Serologic identification and characterization of a macaque T-lymphotropic retrovirus closely related to HTLV-III. *Science*. Jun 7 1985;228(4704):1199-1201.

146. Popovic M, Sarngadharan MG, Read E, Gallo RC. Detection, isolation, and continuous production of cytopathic retroviruses (HTLV-III) from patients with AIDS and pre-AIDS. *Science*. May 4 1984;224(4648):497-500.

147. Sharer LR, Cho ES, Epstein LG. Multinucleated giant cells and HTLV-III in AIDS encephalopathy. *Hum Pathol*. Aug 1985;16(8):760.

148. Dowsett AB, Roff MA, Greenaway PJ, Elphick ER, Farrar GH. Syncytia--a major site for the production of the human immunodeficiency virus? *Aids*. Sep 1987;1(3):147-150.

149. Letvin NL, Daniel MD, Sehgal PK, et al. Induction of AIDS-like disease in macaque monkeys with T-cell tropic retrovirus STLV-III. *Science*. Oct 4 1985;230(4721):71-73.

150. Letvin NL, Hunt RD. An acquired immune deficiency syndrome of macaque monkeys. *Ann N Y Acad Sci*. 1984;437:121-130.

151. Chopra HC, Mason MM. A new virus in a spontaneous mammary tumor of a rhesus monkey. *Cancer Res*. Aug 1970;30(8):2081-2086.

152. Jensen EM, Zelljadt I, Chopra HC, Mason MM. Isolation and propagation of a virus from a spontaneous mammary carcinoma of a rhesus monkey. *Cancer Res*. Sep 1970;30(9):2388-2393.

153. Marx PA, Bryant ML, Osborn KG, et al. Isolation of a new serotype of simian acquired immune deficiency syndrome type D retrovirus from Celebes black macaques (Macaca nigra) with immune deficiency and retroperitoneal fibromatosis. *J Virol*. Nov 1985;56(2):571-578.

154. Keele BF, Jones JH, Terio KA, et al. Increased mortality and AIDS-like immunopathology in wild chimpanzees infected with SIVcpz. *Nature*. Jul 23 2009;460(7254):515-519.

155. Stromberg K, Benveniste RE, Arthur LO, et al. Characterization of exogenous type D retrovirus from a fibroma of a macaque with simian AIDS and fibromatosis. *Science*. Apr 20 1984;224(4646):289-282.

156. Gravell M, London WT, Houff SA, et al. Transmission of simian acquired immunodeficiency syndrome (SAIDS) with blood or filtered plasma. *Science*. Jan 6 1984;223(4631):74-76.

157. Thayer RM, Power MD, Bryant ML, Gardner MB, Barr PJ, Luciw PA. Sequence relationships of type D retroviruses which cause simian acquired immunodeficiency syndrome. *Virology*. Apr 1987;157(2):317-329.

158. Lowenstine LJ, Pedersen NC, Higgins J, et al. Seroepidemiologic survey of captive Old-World primates for antibodies to human and simian retroviruses, and isolation of a lentivirus from sooty mangabeys (Cercocebus atys). *Int J Cancer*. Oct 15 1986;38(4):563-574.

159. Marx PA, Maul DH, Osborn KG, et al. Simian AIDS: isolation of a type D retrovirus and transmission of the disease. *Science*. Mar 9 1984;223(4640):1083-1086.

160. Power MD, Marx PA, Bryant ML, Gardner MB, Barr PJ, Luciw PA. Nucleotide sequence of SRV-1, a type D simian acquired immune deficiency syndrome retrovirus. *Science*. Mar 28 1986;231(4745):1567-1572.

161. Sonigo P, Barker C, Hunter E, Wain-Hobson S. Nucleotide sequence of Mason-Pfizer monkey virus: an immunosuppressive D-type retrovirus. *Cell*. May 9 1986;45(3):375-385.

162. Marx PA, Pedersen NC, Lerche NW, et al. Prevention of simian acquired immune deficiency syndrome with a formalin-inactivated type D retrovirus vaccine. *J Virol*. Nov 1986;60(2):431-435.

163. Press A. AIDS-Related Virus Suppressed. *New York Times.* September 3, 1986, 1986.

164. Gonda MA. Molecular genetics and structure of the human immunodeficiency virus. *J Electron Microsc Tech.* Jan 1988;8(1):17-40.

165. Gonda MA, Braun MJ, Carter SG, et al. Characterization and molecular cloning of a bovine lentivirus related to human immunodeficiency virus. *Nature.* Nov 26-Dec 2 1987;330(6146):388-391.

166. Gonda MA, Braun MJ, Clements JE, et al. Human T-cell lymphotropic virus type III shares sequence homology with a family of pathogenic lentiviruses. *Proc Natl Acad Sci U S A.* Jun 1986;83(11):4007-4011.

167. Stephens RM, Casey JW, Rice NR. Equine infectious anemia virus gag and pol genes: relatedness to visna and AIDS virus. *Science.* Feb 7 1986;231(4738):589-594.

168. Braun MJ, Clements JE, Gonda MA. The visna virus genome: evidence for a hypervariable site in the env gene and sequence homology among lentivirus envelope proteins. *J Virol.* Dec 1987;61(12):4046-4054.

169. Thormar H. Maedi-visna virus and its relationship to human immunodeficiency virus. *AIDS Rev.* Oct-Dec 2005;7(4):233-245.

170. Davis JL, Molineaux S, Clements JE. Visna virus exhibits a complex transcriptional pattern: one aspect of gene expression shared with the acquired immunodeficiency syndrome retrovirus. *J Virol.* May 1987;61(5):1325-1331.

171. Guyader M, Emerman M, Sonigo P, Clavel F, Montagnier L, Alizon M. Genome organization and transactivation of the human immunodeficiency virus type 2. *Nature.* Apr 16-22 1987;326(6114):662-669.

172. St-Louis MC, Cojocariu M, Archambault D. The molecular biology of bovine immunodeficiency virus: a comparison with other lentiviruses. *Anim Health Res Rev.* Dec 2004;5(2):125-143.

173. WHO. Animal models for HIV infection and AIDS: memorandum from a WHO meeting. *Bull World Health Organ.* 1988;66(5):561-574.

174. Bendinelli M, Matteucci D, Friedman H. Retrovirus-induced acquired immunodeficiencies. *Adv Cancer Res.* 1985;45:125-181.

175. Clavel F, Guetard D, Brun-Vezinet F, et al. Isolation of a new human retrovirus from West African patients with AIDS. *Science.* Jul 18 1986;233(4761):343-346.

176. Clavel F, Guyader M, Guetard D, Salle M, Montagnier L, Alizon M. Molecular cloning and polymorphism of the human immune deficiency virus type 2. *Nature.* Dec 18-31 1986;324(6098):691-695.

177. HIV Origin: 'A Continuing Mystery:' Green Monkey Theory Disputed. *Skin & Allergy News.* Vol 19. Bethesda, MD, USA: Charles J. Siegel; 1988:28-29.

178. McNeil Jr. D. The World; AIDS in Africa: The Silent Stalker. *New York Times,* December 27, 1998.

179. Peeters M, Honore C, Huet T, et al. Isolation and partial characterization of an HIV-related virus occurring naturally in chimpanzees in Gabon. *AIDS.* Oct 1989;3(10):625-630.

180. Peeters M, Fransen K, Delaporte E, et al. Isolation and characterization of a new chimpanzee lentivirus (simian immunodeficiency virus isolate cpz-ant) from a wild-captured chimpanzee. *AIDS.* May 1992;6(5):447-451.

181. Huet T, Cheynier R, Meyerhans A, Roelants G, Wain-Hobson S. Genetic organization of a chimpanzee lentivirus related to HIV-1. *Nature.* May 24 1990;345(6273):356-359.

182. Janssens W, Fransen K, Peeters M, et al. Phylogenetic analysis of a new chimpanzee lentivirus SIVcpz-gab2 from a wild-captured chimpanzee from Gabon. *AIDS Res Hum Retroviruses.* Sep 1994;10(9):1191-1192.

183. Vanden Haesevelde MM, Peeters M, Jannes G, et al. Sequence analysis of a highly divergent HIV-1-related lentivirus isolated from a wild captured chimpanzee. *Virology.* Jul 15 1996;221(2):346-350.

184. Gilden RV, Arthur LO, Robey WG, Kelliher JC, Graham CE, Fischinger PJ. HTLV-III antibody in a breeding chimpanzee not experimentally exposed to the virus. *Lancet.* Mar 22 1986;1(8482):678-679.

185. Gajdusek DC, Amyx HL, Gibbs CJ, Jr., et al. Transmission experiments with human T-lymphotropic retroviruses and human AIDS tissue. *Lancet.* Jun 23 1984;1(8391):1415-1416.

186. Alter HJ, Eichberg JW, Masur H, et al. Transmission of HTLV-III infection from human plasma to chimpanzees: an animal model for AIDS. *Science.* Nov 2 1984;226(4674):549-552.

187. Gajdusek DC, Amyx HL, Gibbs CJ, Jr., et al. Infection of chimpanzees by human T-lymphotropic retroviruses in brain and other tissues from AIDS patients. *Lancet.* Jan 5 1985;1(8419):55-56.

188. Fultz PN, McClure HM, Swenson RB, et al. Persistent infection of chimpanzees with human T-lymphotropic virus type III/lymphadenopathy-associated virus: a potential model for acquired immunodeficiency syndrome. *J Virol.* Apr 1986;58(1):116-124.

189. Nara P, Hatch W, Kessler J, Kelliher J, Carter S. The biology of human immunodeficiency virus-1 IIIB infection in the chimpanzee: in vivo and in vitro correlations. *J Med Primatol.* 1989;18(3-4):343-355.

190. Altman KA. H.I.V. Is Linked To a Subspecies Of Chimpanzee. *New York Times.* February 1, 1999, 1999: Front Page.

191. Nelson S. Aids origin 'discovered' BBC; February 1, 1999.

192. Sharp PM, Shaw GM, Hahn BH. Simian immunodeficiency virus infection of chimpanzees. *J Virol.* Apr 2005;79(7):3891-3902.

193. Keele BF, Van Heuverswyn F, Li Y, et al. Chimpanzee reservoirs of pandemic and nonpandemic HIV-1. *Science.* Jul 28 2006;313(5786):523-526.

194. BBC. HIV origin 'found in wild chimps': BBC; May 25, 2006.

195. Corbet S, Muller-Trutwin MC, Versmisse P, et al. env sequences of simian immunodeficiency viruses from chimpanzees in Cameroon are strongly related to those of human immunodeficiency virus group N from the same geographic area. *J Virol.* Jan 2000;74(1):529-534.

196. WHO. Global AIDS surveillance--Part I. *Wkly Epidemiol Rec.* Nov 26 1999;74(47):401-404.

197. UN. *Population in 1990 and 2000: All Countries*: Population Division, Department of Economic and Social Affairs, United Nations.

198. WHO. Acquired Immunodeficiency Syndrome (AIDS): WHO/CDC case definition for AIDS. *Wkly Epidemiol Rec.* March 7 1986;61(10):69-73.

199. WHO. Workshop on AIDS in Central Africa. Paper presented at: Workshop on AIDS in Central Africa; October 22-25, 1985, 1985; Banqui, Central African Republic.

200. Wilson B. India Lowers HIV Estimate; Part of Wider Trend. *Morning Edition.* Washington DC: National Public Radio; July 6, 2007.

201. UNAIDS. Fact Sheet: Revised HIV Estimates. http://data.unaids.org/pub/EPISlides/2007/071118_epi_revisions_factsheet_en.pdf.

202. UN. *Statistical Yearbook, Issue 53.* Geneva; 2010.

203. Clumeck N, Mascart-Lemone F, de Maubeuge J, Brenez D, Marcelis L. Acquired immune deficiency syndrome in Black Africans. *Lancet.* Mar 19 1983;1(8325):642.

204. Clumeck N, Sonnet J, Taelman H, Cran S, Henrivaux P. Acquired immune deficiency syndrome in Belgium and its relation to Central Africa. *Ann N Y Acad Sci.* 1984;437:264-269.

205. Colebunders R, Taelman H, Piot P. AIDS: an old disease from Africa? *Br Med J (Clin Res Ed).* Sep 22 1984;289(6447):765.

206. Glauser MP, Francioli P. Clinical and epidemiological survey of acquired immune deficiency syndrome in Europe. *Eur J Clin Microbiol.* Feb 1984;3(1):55-58.

207. Abrams DI, Lewis BJ, Volberding PA. Lymphadenopathy: endpoint or prodrome? Update of a 24-month prospective study. *Ann N Y Acad Sci.* 1984;437:207-215.

208. Otu AA. Kaposi's sarcoma and HTLV-III: a study in Nigerian adult males. *J R Soc Med.* Sep 1986;79(9):510-514.

209. CDC. Update: acquired immunodeficiency syndrome--Europe. *MMWR Morb Mortal Wkly Rep.* Nov 2 1984;33(43):607-609.

210. Malan R. AIDS In Africa: In Search of the Truth. *Rolling Stone*; November 22, 2001:70, 71, 72, 74, 75, 76, 77, 78, 80, 82, 100, 102.

211. Saxinger C, Gallo RC. Methods in laboratory investigation. Application of the indirect enzyme-linked immunosorbent assay microtest to the detection and surveillance of human t cell leukemia-lymphoma virus. *Lab Invest.* Sep 1983;49(3):371-377.

212. Sarngadharan MG, Popovic M, Bruch L, Schupbach J, Gallo RC. Antibodies reactive with human T-lymphotropic retroviruses (HTLV-III) in the serum of patients with AIDS. *Science.* May 4 1984;224(4648):506-508.

213. Biggar RJ, Melbye M, Kestens L, et al. Seroepidemiology of HTLV-III antibodies in a remote population of eastern Zaire. *Br Med J (Clin Res Ed).* Mar 16 1985;290(6471):808-810.

214. Biggar RJ, Melbye M, Kestems L, et al. Kaposi's sarcoma in Zaire is not associated with HTLV-III infection. *N Engl J Med.* Oct 18 1984;311(16):1051-1052.

215. Altman KA. AIDS In Africa: A Pattern of Mystery *New York Times.* November 8, 1985 1985;World

216. Altman KA. More Data Found on AIDS in Africa. *New York Times.* December 15, 1985, 1985.

217. Schupbach J, Tanner M. Specificity of human immunodeficiency virus (LAV/HTLV-III)-reactive antibodies in African sera from southeastern Tanzania. *Acta Trop.* Sep 1986;43(3):195-206.

218. Clumeck N, Robert-Guroff M, Van de Perre P, et al. Seroepidemiological studies of HTLV-III antibody prevalence among selected groups of heterosexual Africans. *Jama.* Nov 8 1985;254(18):2599-2602.

219. Bayley AC, Downing RG, Cheingsong-Popov R, Tedder RS, Dalgleish AG, Weiss RA. HTLV-III serology distinguishes atypical and endemic Kaposi's sarcoma in Africa. *Lancet.* Feb 16 1985;1(8425):359-361.

220. Carswell JW, Sewankambo N, Lloyd G, Downing RG. How long has the AIDS virus been in Uganda? *Lancet.* May 24 1986;1(8491):1217.

221. Mann JM, Nzilambi N, Piot P, et al. HIV infection and associated risk factors in female prostitutes in Kinshasa, Zaire. *Aids.* Aug 1988;2(4):249-254.

222. Saxinger C, Levine PH, Dean A, Lange-Wantzin G, Gallo R. Unique pattern of HTLV-III (AIDS-related) antigen recognition by sera from African children in Uganda (1972). *Cancer Res.* Sep 1985;45(9 Suppl):4624s-4626s.

223. Mann JM, Francis H, Quinn T, et al. Surveillance for AIDS in a central African city. Kinshasa, Zaire. *Jama.* Jun 20 1986;255(23):3255-3259.

224. Dube DK, Dube S, Erensoy S, et al. Serological and nucleic acid analyses for HIV and HTLV infection on archival human plasma samples from Zaire. *Virology*. Jul 1994;202(1):379-389.

225. FDA. Complete List of Donor Screening Assays for Infectious Agents and HIV Diagnostic Assays. http://www.fda.gov/BiologicsBloodVaccines/BloodBloodProducts/ApprovedProducts/LicensedProductsBLAs/BloodDonorScreening/InfectiousDisease/UCM080466. Accessed October 5, 2010.

226. Fleming AF. Seroepidemiology of human immunodeficiency viruses in Africa. *Biomed Pharmacother*. 1988;42(5):309-320.

227. Hunsmann G, Schneider J, Wendler I, Fleming AF. HTLV positivity in Africans. *Lancet*. Oct 26 1985;2(8461):952-953.

228. Wendler I, Schneider J, Gras B, Fleming AF, Hunsmann G, Schmitz H. Seroepidemiology of human immunodeficiency virus in Africa. *Br Med J (Clin Res Ed)*. Sep 27 1986;293(6550):782-785.

229. Fleury HJ, Babin M, Bonnici JF, Bailly C, Chancerel B, Le Bras M. Virus related to but not identical with LAV/HTLV-III in Cameroon. *Lancet*. Apr 12 1986;1(8485):854.

230. Brun-Vezinet F, Jaeger G, Rouzioux C, et al. Lack of evidence for human or simian T-lymphotropic viruses type III infection in pygmies. *Lancet*. Apr 12 1986;1(8485):854.

231. Okpara RA, Williams E, Schneider J, Wendler I, Hunsmann G. Antibodies to human T-cell leukemia virus types I and III in blood donors from Calabar, Nigeria. *Ann Intern Med*. Jan 1986;104(1):132.

232. Levy JA, Pan LZ, Beth-Giraldo E, et al. Absence of antibodies to the human immunodeficiency virus in sera from Africa prior to 1975. *Proc Natl Acad Sci U S A*. Oct 1986;83(20):7935-7937.

233. Sher R, Antunes S, Reid B, Falcke H. Seroepidemiology of human immunodeficiency virus in Africa from 1970 to 1974. *N Engl J Med*. Aug 13 1987;317(7):450-451.

234. Okpara RA, Akinsete I, Williams EE, Schneider J, Wendler I, Hunsmann G. Antibodies to human immunodeficiency virus (HTLV-III/LAV) in people from Lagos and Cross River States of Nigeria. *Acta Haematol*. 1988;79(2):91-93.

235. Proffitt MR, Yen-Lieberman B. Laboratory diagnosis of human immunodeficiency virus infection. *Infect Dis Clin North Am*. Jun 1993;7(2):203-219.

236. Cordes RJ, Ryan ME. Pitfalls in HIV testing. Application and limitations of current tests. *Postgrad Med*. Nov 1995;98(5):177-180, 185-176, 189.

237. Smith DM, Dewhurst S, Shepherd S, Volsky DJ, Goldsmith JC. False-positive enzyme-linked immunosorbent assay reactions for antibody to human immunodeficiency virus in a population of midwestern patients with congenital bleeding disorders. *Transfusion*. Jan-Feb 1987;27(1):112.

238. Weber B, Moshtaghi-Boronjeni M, Brunner M, Preiser W, Breiner M, Doerr HW. Evaluation of the reliability of 6 current anti-HIV-1/HIV-2 enzyme immunoassays. *J Virol Methods*. Sep 1995;55(1):97-104.

239. Steckelberg JM, Cockerill FR, 3rd. Serologic testing for human immunodeficiency virus antibodies. *Mayo Clin Proc*. Apr 1988;63(4):373-380.

240. Novick DM, Des Jarlais DC, Kreek MJ, et al. Specificity of antibody tests for human immunodeficiency virus in alcohol and parenteral drug abusers with chronic liver disease. *Alcohol Clin Exp Res*. Oct 1988;12(5):687-690.

241. Ng VL. Serological diagnosis with recombinant peptides/proteins. *Clin Chem*. Oct 1991;37(10 Pt 1):1667-1668.

242. Bylund DJ, Ziegner UH, Hooper DG. Review of testing for human immunodeficiency virus. *Clin Lab Med*. Jun 1992;12(2):305-333.

243. Mendenhall CL, Roselle GA, Grossman CJ, Rouster SD, Weesner RE. False positive tests for HTLV-III antibodies in alcoholic patients with hepatitis. *N Engl J Med.* Apr 3 1986;314(14):921-922.

244. Jackson JB, Balfour HH, Jr. Practical diagnostic testing for human immunodeficiency virus. *Clin Microbiol Rev.* Jan 1988;1(1):124-138.

245. Celum CL, Coombs RW, Jones M, et al. Risk factors for repeatedly reactive HIV-1 EIA and indeterminate western blots. A population-based case-control study. *Arch Intern Med.* May 23 1994;154(10):1129-1137.

246. Volsky DJ, Rodriguez L, Dewhurst S, et al. Antibodies to acquired immune deficiency syndrome (AIDS)-associated virus (HTLV-III/LAV) in Venezuelan populations. *AIDS Res.* Spring 1986;2(2):79-92.

247. Nahmias AJ, Weiss J, Yao X, et al. Evidence for human infection with an HTLV III/LAV-like virus in Central Africa, 1959. *Lancet.* May 31 1986;1(8492):1279-1280.

248. Burke DS. Laboratory diagnosis of human immunodeficiency virus infection. *Clin Lab Med.* Sep 1989;9(3):369-392.

249. Garrett AJ, Seagroatt V, Supran EM, Habermehl KO, Hampl H, Schild GC. Measurement of antibodies to human immunodeficiency virus: an international collaborative study to evaluate WHO reference sera. *Bull World Health Organ.* 1988;66(2):197-202.

250. Nuttall P, Pratt R, Nuttall L, Daly C. False-positive results with HIV ELISA kits. *Lancet.* Aug 30 1986;2(8505):512-513.

251. Lai-Goldman M, McBride JH, Howanitz PJ, Rodgerson DO, Miles JA, Peter JB. Presence of HTLV-III antibodies in immune serum globulin preparations. *Am J Clin Pathol.* May 1987;87(5):635-639.

252. Gentili G, Wirz M, Puccinelli M, Mele C, Collotti C, Vicari G. Detection of anti-HIV antibodies in immunoglobulin preparations: the significance of antibodies to the HIV-envelope. *Biologicals.* Jul 1991;19(3):197-202.

253. Gocke DJ, Raska K, Jr., Pollack W, Schwartzer T. HTLV-III antibody in commercial immunoglobulin. *Lancet.* Jan 4 1986;1(8471):37-38.

254. Tedder RS, Uttley A, Cheingsong-Popov R. Safety of immunoglobulin preparation containing anti-HTLV-III. *Lancet.* Apr 6 1985;1(8432):815.

255. Wood CC, Williams AE, McNamara JG, Annunziata JA, Feorino PM, Conway CO. Antibody against the human immunodeficiency virus in commercial intravenous gammaglobulin preparations. *Ann Intern Med.* Oct 1986;105(4):536-538.

256. WHO. Joint United Nations Programme on HIV/AIDS (UNAIDS)-WHO. Revised recommendations for the selection and use of HIV antibody tests. *Wkly Epidemiol Rec.* Mar 21 1997;72(12):81-87.

257. Houn HY, Pappas AA, Walker EM, Jr. Status of current clinical tests for human immunodeficiency virus (HIV): applications and limitations. *Ann Clin Lab Sci.* Sep-Oct 1987;17(5):279-285.

258. Biggar RJ, Gigase PL, Melbye M, et al. ELISA HTLV retrovirus antibody reactivity associated with malaria and immune complexes in healthy Africans. *Lancet.* Sep 7 1985;2(8454):520-523.

259. Andrade VL, Avelleira JC, Marques A, Vianna FR, Schechter M. Leprosy as cause of false-positive results in serological assays for the detection of antibodies to HIV-1. *Int J Lepr Other Mycobact Dis.* Mar 1991;59(1):125-126.

260. Cann AJ, Karn J. Molecular biology of HIV: new insights into the virus life-cycle. *AIDS.* 1989;3 Suppl 1:S19-34.

261. WHO. Acquired immunodeficiency syndrome (AIDS). Proposed WHO criteria for interpreting results from western blot assays for HIV-1, HIV-2, and HTLV-I/HTLV-II. *Wkly Epidemiol Rec.* Sep 14 1990;65(37):281-283.

262. Meyer KB, Pauker SG. Screening for HIV: can we afford the false positive rate? *N Engl J Med.* Jul 23 1987;317(4):238-241.

263. Burke DS, Redfield RR. False-positive Western blot tests for antibodies to HTLV-III. *JAMA.* Jul 18 1986;256(3):347.

264. Biberfeld G, Bredberg-Raden U, Bottiger B, et al. Blood donor sera with false-positive Western blot reactions to human immunodeficiency virus. *Lancet.* Aug 2 1986;2(8501):289-290.

265. CDC. Interpretation and use of the western blot assay for serodiagnosis of human immunodeficiency virus type 1 infections. *MMWR Morb Mortal Wkly Rep.* Jul 21 1989;38(Suppl 7):1-7.

266. Robey WG, Safai B, Oroszlan S, et al. Characterization of envelope and core structural gene products of HTLV-III with sera from AIDS patients. *Science.* May 3 1985;228(4699):593-595.

267. Saxinger WC, Levine PH, Dean AG, et al. Evidence for exposure to HTLV-III in Uganda before 1973. *Science.* Mar 1 1985;227(4690):1036-1038.

268. Doran TI, Parra E. False-positive and indeterminate human immunodeficiency virus test results in pregnant women. *Arch Fam Med.* Sep-Oct 2000;9(9):924-929.

269. Sayre KR, Dodd RY, Tegtmeier G, Layug L, Alexander SS, Busch MP. False-positive human immunodeficiency virus type 1 western blot tests in noninfected blood donors. *Transfusion.* Jan 1996;36(1):45-52.

270. De Cock KM, Brun-Vezinet F. Epidemiology of HIV-2 infection. *AIDS.* 1989;3 Suppl 1:S89-95.

271. Kenny DF, Garsia RJ, Gatenby PA, Basten A. Identification of biological false positives in anti-HIV antibody tests. *Aids.* May 1987;1(1):63-64.

272. Hunter JB, Menitove JE. HLA antibodies detected by ELISA HTLV-III antibody kits. *Lancet.* Aug 17 1985;2(8451):397.

273. Sayers MH, Beatty PG, Hansen JA. HLA antibodies as a cause of false-positive reactions in screening enzyme immunoassays for antibodies to human T-lymphotropic virus type III. *Transfusion.* Jan-Feb 1986;26(1):113-115.

274. Blanton M, Balakrishnan K, Dumaswala U, Zelenski K, Greenwalt TJ. HLA antibodies in blood donors with reactive screening tests for antibody to the immunodeficiency virus. *Transfusion.* Jan-Feb 1987;27(1):118-119.

275. Thorpe R, Bird C, Garrett AJ, Minor PD, Schild GC, Thomas DP. False-positive immunoblot results with antibodies against human immunodeficiency virus. *Lancet.* Sep 13 1986;2(8507):627-628.

276. Shacham E, Reece M, Ong'or WO, Omollo O, Basta TB. A cross-cultural comparison of psychological distress among individuals living with HIV in Atlanta, Georgia, and Eldoret, Kenya. *J Int Assoc Physicians AIDS Care (Chic Ill).* May-Jun 2010;9(3):162-169.

277. Ameglio F, Benedetto A, Marotta P, et al. A high proportion of sera of heroin addicts possesses anti-HLA class I and class II reactivity. *Clin Immunol Immunopathol.* Feb 1988;46(2):328-334.

278. Ameglio F, Dolei A, Benedetto A, Sorrentino R, Tanigaki N, Tosi R. Antibodies reactive with nonpolymorphic epitopes on HLA molecules interfere in screening tests for the human immunodeficiency virus. *J Infect Dis.* Dec 1987;156(6):1034-1035.

279. Ameglio F, Saba F, Bitti A, et al. Antibody reactivity to HLA classes I and II in sera from patients with hydatidosis. *J Infect Dis.* Oct 1987;156(4):673-676.

280. Voevodin A. HIV screening in Russia. *Lancet.* Jun 20 1992;339(8808):1548.

281. Shima-Sano T, Yamada R, Sekita K, et al. A human immunodeficiency virus screening algorithm to address the high rate of false-positive results in pregnant women in Japan. *PLoS One.* 2010;5(2):e9382.

282. Tung CS, Sangi-Haghpeykar H, Levison J. Rapid versus standard testing for prenatal HIV screening in a predominantly Hispanic population. *J Perinatol.* Jan 2010;30(1):30-32.

283. WHO. *Regional workshop on HIV testing strategies.* Amman, Jordan: World Health Organization Regional Office for the Eastern Mediterranean; June 26, 2006 2006.

284. Salomon J, Gakidou E, Murray J. *Methods for Modeling the HIV/AIDS Epidemic In Sub-Saharan Africa*: UNAIDS/WHO Working Group on Global HIV/AIDS and STD Surveillance.

285. Diaz T, De Cock K, Brown T, Ghys PD, Boerma JT. New strategies for HIV surveillance in resource-constrained settings: an overview. *AIDS.* May 2005;19 Suppl 2:S1-8.

286. Quinn TC, Mann JM, Curran JW, Piot P. AIDS in Africa: an epidemiologic paradigm. *Science.* Nov 21 1986;234(4779):955-963.

287. Fleming AF. AIDS in Africa--an update. *AIDS Forsch.* Mar 1988;3(3):116-138.

288. Derbyshire SW. WHO criticised for "inflating" AIDS figures. *AIDS Anal Afr.* Dec 1995;5(6):4-5.

289. Mid-year population estimates 2010. In: Africa SS, ed. Pretoria, South Africa; 2010.

290. Makubalo L, Netshidzivhani P, Mahlasela L, du Plessis R. *National HIV and Syphilis Antenatal Sero-Prevalence Survey in South Africa 2003* 2003.

291. Mortality and causes of death in South Africa, 1997-2003: Findings from death notifications. In: Africa SS, ed. Pretoria, South Africa: Statistics South Africa; 2005.

292. Mortality and causes of death in South Africa, 2008: Findings from death notification. In: Africa SS, ed. Pretoria, South Africa: Statistics South Africa, Pretoria; 2010.

293. Mortality and causes of death in South Africa, 2003 and 2004 Findings from death notification. In: Africa SS, ed. Pretoria, South Africa: Statistics South Africa; 2006.

294. Mortality and causes of death in South Africa, 2006: Findings from death notification. In: Africa SS, ed. Pretoria, South Africa: Statistics South Africa, Pretoria 2008.

295. Mortality and causes of death in South Africa, 2007: Findings from death notification. In: Africa SS, ed. Pretoria, South Africa: Statistics South Africa, Pretoria 2009.

296. Mortality and causes of death in South Africa, 2005: Findings from death notification. In: Africa SS, ed. Pretoria, South Africa: Statistics South Africa, Pretoria; 2007.

297. CLSI. *How to Define, Determine, and Utilize Reference Intervals in the Clinical Laboratory; Approved Guideline (NCCLS Document C28-A2).* Second Edition ed. Wayne, PA: Clinical and Laboratory Standards Institute; 2000.

298. Martin PW, Burger DR, Caouette S, Goldstein AS, Peetoom F. Importance of confirmatory tests after strongly positive HTLV-III screening tests. *N Engl J Med.* Jun 12 1986;314(24):1577.

299. Greenberg AE, Schable CA, Sulzer AJ, Collins WE, Phuc N-D. Evaluation of serological cross-reactivity between antibodies to Plasmodium and HTLV-III/LAV. *Lancet.* Aug 2 1986;2(8501):247-249.

300. Biggar RJ, Johnson BK, Oster C, et al. Regional variation in prevalence of antibody against human T-lymphotropic virus types I and III in Kenya, East Africa. *Int J Cancer.* Jun 15 1985;35(6):763-767.

301. Grunnet N, Jersild C, Georgsen J. Photometric reading of anti-HTLV-III ELISA kits. *Lancet.* Dec 7 1985;2(8467):1302.

302. Volsky DJ, Wu YT, Stevenson M, et al. Antibodies to HTLV-III/LAV in Venezuelan patients with acute malarial infections. *N Engl J Med.* Mar 6 1986;314(10):647-648.

303. Ronalds CJ, Grint PC, Hardiman AE. Anti-HIV testing on urgent specimens. *Lancet.* Feb 7 1987;1(8528):323-324.

304. Denis F, Barin F, Gershy-Damet G, et al. Prevalence of human T-lymphotropic retroviruses type III (HIV) and type IV in Ivory Coast. *Lancet.* Feb 21 1987;1(8530):408-411.

305. Van de Perre P, Clumeck N, Steens M, et al. Seroepidemiological study on sexually transmitted diseases and hepatitis B in African promiscuous heterosexuals in relation to HTLV-III infection. *Eur J Epidemiol.* Mar 1987;3(1):14-18.

306. Piot P, Plummer FA, Rey MA, et al. Retrospective seroepidemiology of AIDS virus infection in Nairobi populations. *J Infect Dis.* Jun 1987;155(6):1108-1112.

307. N'Galy B, Ryder RW. Epidemiology of HIV infection in Africa. *J Acquir Immune Defic Syndr.* 1988;1(6):551-558.

308. Mabey DC, Tedder RS, Hughes AS, et al. Human retroviral infections in The Gambia: prevalence and clinical features. *Br Med J (Clin Res Ed).* Jan 9 1988;296(6615):83-86.

309. Nzilambi N, De Cock KM, Forthal DN, et al. The prevalence of infection with human immunodeficiency virus over a 10-year period in rural Zaire. *N Engl J Med.* Feb 4 1988;318(5):276-279.

310. Marlink RG, Ricard D, M'Boup S, et al. Clinical, hematologic, and immunologic cross-sectional evaluation of individuals exposed to human immunodeficiency virus type-2 (HIV-2). *AIDS Res Hum Retroviruses.* Apr 1988;4(2):137-148.

311. Chiphangwi J, Liomba G, Ntaba HM, et al. Human immunodeficiency virus infection is prevalent in Malawi. *Infection.* 1987;15(5):363.

312. Van de Perre P, Clumeck N, Carael M, et al. Female prostitutes: a risk group for infection with human T-cell lymphotropic virus type III. *Lancet.* Sep 7 1985;2(8454):524-527.

313. Kreiss JK, Koech D, Plummer FA, et al. AIDS virus infection in Nairobi prostitutes. Spread of the epidemic to East Africa. *N Engl J Med.* Feb 13 1986;314(7):414-418.

314. Chikwem JO, Ola TO, Gashau W, Chikwem SD, Bajami M, Mambula S. Impact of health education on prostitutes' awareness and attitudes to acquired immune deficiency syndrome (AIDS). *Public Health.* Sep 1988;102(5):439-445.

315. Kpatinde F. [Prostitution and AIDS. The party continues, the scourge advances]. *Jeune Afr.* Dec 2 1987(1404):52-55.

316. Yeboah-afari A. Helping prostitutes in Accra. *AIDS Watch.* 1988(4):4-5.

317. Ngugi EN, Plummer FA, Simonsen JN, et al. Prevention of transmission of human immunodeficiency virus in Africa: effectiveness of condom promotion and health education among prostitutes. *Lancet.* Oct 15 1988;2(8616):887-890.

318. Bchir A, Jemni L, Saadi M, Milovanovic A, Brahim H, Catalan F. Markers of sexually transmitted diseases in prostitutes in central Tunisia. *Genitourin Med.* Dec 1988;64(6):396-397.

319. Day S. Prostitute women and AIDS: anthropology. *AIDS.* Dec 1988;2(6):421-428.

320. Mann J, Quinn TC, Piot P, et al. Condom use and HIV infection among prostitutes in Zaire. *N Engl J Med.* Feb 5 1987;316(6):345.

321. Piot P, Laga M. Prostitutes: a high risk group for HIV infection? *Soz Praventivmed.* 1988;33(7):336-339.

322. Elm J, Desowitz R, Diwan A. Serological cross-reactivities between the retroviruses HIV and HTLV-1 and the malaria parasite Plasmodium falciparum. *P N G Med J.* Mar 1998;41(1):15-22.

323. Rodriquez L, Dewhurst S, Sinangil F, Merino F, Godoy G, Volsky DJ. Antibodies to HTLV-III/LAV among aboriginal Amazonian Indians in Venezuela. *Lancet.* Nov 16 1985;2(8464):1098-1100.

324. Mohammed I, Nasidi A, Chikwem JO, et al. HIV infection in Nigeria. *Aids.* Feb 1988;2(1):61-62.

325. Jacobs RM, Smith HE, Gregory B, Valli VE, Whetstone CA. Detection of multiple retroviral infections in cattle and cross-reactivity of bovine immunodeficiency-like virus and human immunodeficiency virus type 1 proteins using bovine and human sera in a western blot assay. *Can J Vet Res.* Oct 1992;56(4):353-359.

326. Azocar J, Martinez C, McLane MF, Allan J, Essex M. Lack of endemic HIV infection in Venezuela. *AIDS Res Hum Retroviruses.* Summer 1987;3(2):107-108.

327. Watt G, Chanbancherd P, Brown AE. Human immunodeficiency virus type 1 test results in patients with malaria and dengue infections. *Clin Infect Dis.* May 2000;30(5):819.

328. Parry JV, Richmond J, Edwards N, Noone A. Spurious malarial antibodies in HIV infection. *Lancet.* Dec 5 1992;340(8832):1412-1413.

329. Nguyen-Dinh P, Greenberg AE, Mann JM, et al. Absence of association between Plasmodium falciparum malaria and human immunodeficiency virus infection in children in Kinshasa, Zaire. *Bull World Health Organ.* 1987;65(5):607-613.

330. Facer CA, Bentley A, Withers M, Kataaha PK. Malaria and ELISA HTLV-III antibody reactivity. *Trans R Soc Trop Med Hyg.* 1986;80(2):351-352.

331. Chrystie IL, Palmer SJ, Voller A, Banatvala JE. False-positive serology and HIV infection. *Lancet.* Feb 13 1993;341(8842):441-442.

332. Biggar RJ. Possible nonspecific association between malaria and HTLV-III/LAV. *N Engl J Med.* Aug 14 1986;315(7):457-458.

333. Norman C. Africa and the origin of AIDS. *Science.* Dec 6 1985;230(4730):1141.

334. Charmot G, Simon F. [HIV infection and malaria]. *Rev Prat.* Oct 11 1990;40(23):2141-2143.

335. Greenberg AE, Nguyen-Dinh P, Mann JM, et al. The association between malaria, blood transfusions, and HIV seropositivity in a pediatric population in Kinshasa, Zaire. *JAMA.* Jan 22-29 1988;259(4):545-549.

336. Schneider WH, Drucker E. Blood transfusions in the early years of AIDS in sub-Saharan Africa. *Am J Public Health.* Jun 2006;96(6):984-994.

337. Equine Infectious Anemia: Transmission and Pathogenesis. *The Merck Veterinary Manual* [http://www.merckvetmanual.com/mvm/index.jsp?cfile=htm/bc/52800.htm. Accessed October 22, 2010.

338. Piot P, Plummer FA, Mhalu FS, Lamboray JL, Chin J, Mann JM. AIDS: an international perspective. *Science.* Feb 5 1988;239(4840):573-579.

339. Colebunders R, Mann JM, Francis H, et al. Evaluation of a clinical case-definition of acquired immunodeficiency syndrome in Africa. *Lancet.* Feb 28 1987;1(8531):492-494.

340. Harries A. Some clinical aspects of HIV infection in Africa. *Afr Health.* Jul 1992;14(5):10-11.

341. Bygbjerg IC, Schiodt M, Bakilana PB, et al. Usefulness of a clinical case-definition of AIDS in East Africa. *Lancet.* Sep 5 1987;2(8558):569.

342. Van de Perre P, Nzaramba D, Ntilivamunda A, et al. AIDS definition for Africa. *Lancet.* Jul 11 1987;2(8550):99-100.

343. Strecker W, Gurtler L, Schilling M, Binibangili M, Strecker K. Epidemiology and clinical manifestation of HIV infection in northern Zaire. *Eur J Epidemiol.* Feb 1994;10(1):95-98.

344. Grant AD, Djomand G, De Cock KM. Natural history and spectrum of disease in adults with HIV/AIDS in Africa. *AIDS.* 1997;11 Suppl B:S43-54.

345. Gilks CF. What use is a clinical case definition for AIDS in Africa? *BMJ.* Nov 9 1991;303(6811):1189-1190.

346. Murray JF. Pulmonary complications of HIV-1 infection among adults living in Sub-Saharan Africa. *Int J Tuberc Lung Dis.* Aug 2005;9(8):826-835.

347. Fleming AF. Opportunistic infections in AIDS in developed and developing countries. *Trans R Soc Trop Med Hyg.* 1990;84 Suppl 1:1-6.

348. Lucas SB, De Cock KM, Hounnou A, et al. Contribution of tuberculosis to slim disease in Africa *BMJ.* Jun 11 1994;308(6943):1531-1533.

349. Matondo P. Case definitions for AIDS surveillance in Africa. *BMJ.* Jan 4 1992;304(6818):54.

350. Keou FX, Belec L, Esunge PM, Cancre N, Gresenguet G. World Health Organization clinical case definition for AIDS in Africa: an analysis of evaluations. *East Afr Med J.* Oct 1992;69(10):550-553.

351. Gilks CF, Otieno LS, Brindle RJ, et al. The presentation and outcome of HIV-related disease in Nairobi. *Q J Med.* Jan 1992;82(297):25-32.

352. De Cock KM, Selik RM, Soro B, Gayle H, Colebunders RL. For debate. AIDS surveillance in Africa: a reappraisal of case definitions. *BMJ.* Nov 9 1991;303(6811):1185-1188.

353. Lepage P, van de Perre P, Dabis F, et al. Evaluation and simplification of the World Health Organization clinical case definition for paediatric AIDS. *AIDS.* Apr 1989;3(4):221-225.

354. WHO. WHO case definitions for AIDS surveillance in adults and adolescents. *Wkly Epidemiol Rec.* Sep 16 1994;69(37):273-275.

355. Serwadda D, Mugerwa RD, Sewankambo NK, et al. Slim disease: a new disease in Uganda and its association with HTLV-III infection. *Lancet.* Oct 19 1985;2(8460):849-852.

356. Hira SK, Ngandu N, Wadhawan D, et al. Clinical and epidemiological features of HIV infection at a referral clinic in Zambia. *J Acquir Immune Defic Syndr.* 1990;3(1):87-91.

357. O'Keefe EA, Wood R. AIDS in Africa. *Scand J Gastroenterol Suppl.* 1996;220:147-152.

358. Gilks CF, Brindle RJ, Otieno LS, et al. Extrapulmonary and disseminated tuberculosis in HIV-1-seropositive patients presenting to the acute medical services in Nairobi. *AIDS.* Oct 1990;4(10):981-985.

359. Gilks CF, Ojoo SA, Brindle RJ. Non-opportunistic bacterial infections in HIV-seropositive adults in Nairobi, Kenya. *AIDS.* 1991;5 Suppl 1:S113-116.

360. Gilks CF, Brindle RJ, Otieno LS, et al. Life-threatening bacteraemia in HIV-1 seropositive adults admitted to hospital in Nairobi, Kenya. *Lancet.* Sep 1 1990;336(8714):545-549.

361. Abouya YL, Beaumel A, Lucas S, et al. Pneumocystis carinii pneumonia. An uncommon cause of death in African patients with acquired immunodeficiency syndrome. *Am Rev Respir Dis.* Mar 1992;145(3):617-620.

362. Russian DA, Kovacs JA. Pneumocystis carinii in Africa: an emerging pathogen? *Lancet.* Nov 11 1995;346(8985):1242-1243.

363. Harries A. Some clinical aspects of HIV infection in Africa. *Afr Health.* Jul 1991;13(5):25-26.

364. Carme B, Mboussa J, Andzin M, Mbouni E, Mpele P, Datry A. Pneumocystis carinii is rare in AIDS in Central Africa. *Trans R Soc Trop Med Hyg.* Jan-Feb 1991;85(1):80.

365. Serwadda D, Goodgame R, Lucas S, Kocjan G. Absence of pneumocystosis in Ugandan AIDS patients. *AIDS.* Jan 1989;3(1):47-48.

366. Lucas SB. Missing infections in AIDS. *Trans R Soc Trop Med Hyg.* 1990;84 Suppl 1:34-38.

367. Fisk DT, Meshnick S, Kazanjian PH. Pneumocystis carinii pneumonia in patients in the developing world who have acquired immunodeficiency syndrome. *Clin Infect Dis.* Jan 1 2003;36(1):70-78.

368. Hughes WT. Pneumocystis carinii pneumonia. *N Engl J Med.* Dec 22 1977;297(25):1381-1383.

369. Graham SM. Non-tuberculosis opportunistic infections and other lung diseases in HIV-infected infants and children. *Int J Tuberc Lung Dis.* Jun 2005;9(6):592-602.

370. Wakefield AE, Stewart TJ, Moxon ER, Marsh K, Hopkin JM. Infection with Pneumocystis carinii is prevalent in healthy Gambian children. *Trans R Soc Trop Med Hyg.* Nov-Dec 1990;84(6):800-802.

371. Hughes WT. Pneumocystis carinii pneumonitis. *Chest.* Jun 1984;85(6):810-813.

372. Pifer LL, Hughes WT, Stagno S, Woods D. Pneumocystis carinii infection: evidence for high prevalence in normal and immunosuppressed children. *Pediatrics.* Jan 1978;61(1):35-41.

373. Mahomed AG, Murray J, Klempman S, et al. Pneumocystis carinii pneumonia in HIV infected patients from South Africa. *East Afr Med J.* Feb 1999;76(2):80-84.

374. Chakaya JM, Bii C, Ng'ang'a L, et al. Pneumocystis carinii pneumonia in HIV/AIDS patients at an urban district hospital in Kenya. *East Afr Med J.* Jan 2003;80(1):30-35.

375. Malin AS, Gwanzura LK, Klein S, Robertson VJ, Musvaire P, Mason PR. Pneumocystis carinii pneumonia in Zimbabwe. *Lancet.* Nov 11 1995;346(8985):1258-1261.

376. Ruffini DD, Madhi SA. The high burden of Pneumocystis carinii pneumonia in African HIV-1-infected children hospitalized for severe pneumonia. *AIDS.* Jan 4 2002;16(1):105-112.

377. McLeod DT, Neill P, Gwanzura L, et al. Pneumocystis carinii pneumonia in patients with AIDS in Central Africa. *Respir Med.* May 1990;84(3):225-228.

378. Zar HJ, Hanslo D, Tannenbaum E, et al. Aetiology and outcome of pneumonia in human immunodeficiency virus-infected children hospitalized in South Africa. *Acta Paediatr.* Feb 2001;90(2):119-125.

379. Garcia-Jardon M, Bhat VG, Blanco-Blanco E, Stepian A. Postmortem findings in HIV/AIDS patients in a tertiary care hospital in rural South Africa. *Trop Doct.* Apr 2010;40(2):81-84.

380. CDC. Epidemiologic aspects of the current outbreak of Kaposi's sarcoma and opportunistic infections. *N Engl J Med.* Jan 28 1982;306(4):248-252.

381. Colebunders R, Mann JM, Francis H, et al. Generalized papular pruritic eruption in African patients with human immunodeficiency virus infection. *AIDS.* Jul 1987;1(2):117-121.

382. Liautaud B, Pape JW, DeHovitz JA, et al. Pruritic skin lesions. A common initial presentation of acquired immunodeficiency syndrome. *Arch Dermatol.* May 1989;125(5):629-632.

383. Widy-Wirski R, Berkley S, Downing R, et al. Evaluation of the WHO clinical case definition for AIDS in Uganda. *JAMA.* Dec 9 1988;260(22):3286-3289.

384. O'Brien TR, Kedes D, Ganem D, et al. Evidence for concurrent epidemics of human herpesvirus 8 and human immunodeficiency virus type 1 in US homosexual men: rates, risk factors, and relationship to Kaposi's sarcoma. *J Infect Dis.* Oct 1999;180(4):1010-1017.

385. Schulz TF. Kaposi's sarcoma-associated herpesvirus (human herpesvirus 8): epidemiology and pathogenesis. *J Antimicrob Chemother.* Apr 2000;45 Suppl T3:15-27.

386. Renwick N, Schulz T, Goudsmit J. *Kaposi's Sarcoma and Kaposi's Sarcoma-associated Herpesvirus/Human Herpesvirus 8: An Overview*: Los Alamos National Laboratory, Los Alamos, New Mexico; 1999.

387. O'Leary JJ, Kennedy MM, McGee JO. Kaposi's sarcoma associated herpes virus (KSHV/HHV 8): epidemiology, molecular biology and tissue distribution. *Mol Pathol.* Feb 1997;50(1):4-8.

388. Miles SA. Pathogenesis of human immunodeficiency virus-related Kaposi's sarcoma. *Curr Opin Oncol.* Oct 1992;4(5):875-882.

389. Chang Y, Ziegler J, Wabinga H, et al. Kaposi's sarcoma-associated herpesvirus and Kaposi's sarcoma in Africa. Uganda Kaposi's Sarcoma Study Group. *Arch Intern Med.* Jan 22 1996;156(2):202-204.

390. Nawar E, Mbulaiteye SM, Gallant JE, et al. Risk factors for Kaposi's sarcoma among HHV-8 seropositive homosexual men with AIDS. *Int J Cancer.* Jun 10 2005;115(2):296-300.

391. Mancuso R, Biffi R, Valli M, et al. HHV8 a subtype is associated with rapidly evolving classic Kaposi's sarcoma. *J Med Virol.* Dec 2008;80(12):2153-2160.

392. Matondo P. Kaposi's sarcoma and faecal-oral exposure. *Lancet.* Jun 13 1992;339(8807):1490.

393. Darrow WW, Peterman TA, Jaffe HW, Rogers MF, Curran JW, Beral V. Kaposi's sarcoma and exposure to faeces. *Lancet.* Mar 14 1992;339(8794):685.

394. Elford J, Tindall B, Sharkey T. Kaposi's sarcoma and insertive rimming. *Lancet.* Apr 11 1992;339(8798):938.

395. Page-Bodkin K, Tappero J, Samuel M, Winkelstein W. Kaposi's sarcoma and faecal-oral exposure. *Lancet.* Jun 13 1992;339(8807):1490.

396. O'Leary JJ. Seeking the cause of Kaposi's sarcoma. *Nat Med.* Aug 1996;2(8):862-863.

397. Merck. Tuberculosis (TB). *The Merck Manual for Health Care Professionals* [http://www.merckmanuals.com/professional/infectious_diseases/mycobacteria/tuberculosis_tb.html. Accessed December 31, 2011.

398. Bayley AC. Aggressive Kaposi's sarcoma in Zambia, 1983. *Lancet.* Jun 16 1984;1(8390):1318-1320.

399. Coker R, Wood PB. Changing patterns of Kaposi's sarcoma in N.E. Zaire. *Trans R Soc Trop Med Hyg.* 1986;80(6):965-966.

400. Pica F, Volpi A. Transmission of human herpesvirus 8: an update. *Curr Opin Infect Dis.* Apr 2007;20(2):152-156.

401. Vitale F, Viviano E, Perna AM, et al. Serological and virological evidence of non-sexual transmission of human herpesvirus type 8 (HHV8). *Epidemiol Infect.* Dec 2000;125(3):671-675.

402. Simonart T. Iron: a target for the management of Kaposi's sarcoma? *BMC Cancer.* Jan 15 2004;4:1.

403. Szajerka T, Jablecki J. Kaposi's sarcoma revisited. *AIDS Rev.* Oct-Dec 2007;9(4):230-236.

404. Ravn P, Lundgren JD, Kjaeldgaard P, et al. Nosocomial outbreak of cryptosporidiosis in AIDS patients. *BMJ.* Feb 2 1991;302(6771):277-280.

405. Sewankambo N, Mugerwa RD, Goodgame R, et al. Enteropathic AIDS in Uganda. An endoscopic, histological and microbiological study. *Aids.* May 1987;1(1):9-13.

406. Mhiri C, Belec L, Di Costanzo B, Georges A, Gherardi R. The slim disease in African patients with AIDS. *Trans R Soc Trop Med Hyg.* May-Jun 1992;86(3):303-306.

407. CDC, WHO. Acquired Immunodeficiency Syndrome (AIDS): 1987 revision of CDC/WHO case definition. *Wkly Epidemiol Rec.* January 1988 1988;63(1/2):1-8.

408. Carswell JW. Clinical manifestations of AIDS in tropical countries. *Trop Doct.* Oct 1988;18(4):147-150.

409. AIDS and Africa. *Lancet.* December 6, 1986 1986;328(8519):1348.

410. Carael M, Van de Perre PH, Lepage PH, et al. Human immunodeficiency virus transmission among heterosexual couples in Central Africa. *AIDS.* Jun 1988;2(3):201-205.

411. Padian NS, Shiboski SC, Jewell NP. Female-to-male transmission of human immunodeficiency virus. *JAMA.* Sep 25 1991;266(12):1664-1667.

412. Padian NS, Shiboski SC, Glass SO, Vittinghoff E. Heterosexual transmission of human immunodeficiency virus (HIV) in northern California: results from a ten-year study. *Am J Epidemiol.* Aug 15 1997;146(4):350-357.

413. Comparison of female to male and male to female transmission of HIV in 563 stable couples. European Study Group on Heterosexual Transmission of HIV. *BMJ.* Mar 28 1992;304(6830):809-813.

414. Nicolosi A, Correa Leite ML, Musicco M, Arici C, Gavazzeni G, Lazzarin A. The efficiency of male-to-female and female-to-male sexual transmission of the human immunodeficiency virus: a study of 730 stable couples. Italian Study Group on HIV Heterosexual Transmission. *Epidemiology.* Nov 1994;5(6):570-575.

415. Haverkos HW, Battjes RJ. Female-to-male transmission of HIV. *JAMA.* Oct 14 1992;268(14):1855; author reply 1856-1857.

416. CDC. Revision of the CDC surveillance case definition for acquired immunodeficiency syndrome. Council of State and Territorial Epidemiologists; AIDS Program, Center for Infectious Diseases. *MMWR Morb Mortal Wkly Rep.* Aug 14 1987;36 Suppl 1:1S-15S.

417. Law C. Sexually transmitted diseases and enteric infections in the male homosexual population. *Semin Dermatol.* Jun 1990;9(2):178-184.

418. Petithory JC, Derouin F. AIDS and strongyloidiasis in Africa. *Lancet.* Apr 18 1987;1(8538):921.

419. Oettle AG. Geographical and racial differences in the frequency of Kaposi's sarcoma as evidence of environmental or genetic causes. *Acta Unio Int Contra Cancrum.* 1962;18:330-363.

420. Malope BI, MacPhail P, Mbisa G, et al. No evidence of sexual transmission of Kaposi's sarcoma herpes virus in a heterosexual South African population. *AIDS.* Feb 19 2008;22(4):519-526.

421. Vangroenweghe D. The earliest cases of human immunodeficiency virus type 1 group M in Congo-Kinshasa, Rwanda and Burundi and the origin of acquired immune deficiency syndrome. *Philos Trans R Soc Lond B Biol Sci.* Jun 29 2001;356(1410):923-925.

422. Huminer D, Rosenfeld JB, Pitlik SD. AIDS in the pre-AIDS era. *Rev Infect Dis.* Nov-Dec 1987;9(6):1102-1108.

423. Sonnet J, Michaux JL, Zech F, Brucher JM, de Bruyere M, Burtonboy G. Early AIDS cases originating from Zaire and Burundi (1962-1976). *Scand J Infect Dis.* 1987;19(5):511-517.

424. Williams G, Stretton TB, Leonard JC. AIDS in 1959? *Lancet.* Nov 12 1983;2(8359):1136.

425. Williams G, Stretton TB, Leonard JC. Cytomegalic inclusion disease and Pneumocystis carinii infection in an adult. *Lancet.* Oct 29 1960;2(7157):951-955.

426. Corbitt G, Bailey AS, Williams G. HIV infection in Manchester, 1959. *Lancet.* Jul 7 1990;336(8706):51.

427. Corbitt G, Bailey AS. AIDS in Manchester, 1959? *Lancet.* Apr 22 1995;345(8956):1058.

428. Bailey AS, Corbitt G. Was HIV present in 1959? *Lancet.* Jan 20 1996;347(8995):189.

429. Zhu T, Ho DD. Was HIV present in 1959? *Nature.* Apr 6 1995;374(6522):503-504.

430. Hooper E, Hamilton WD. 1959 Manchester case of syndrome resembling AIDS. *Lancet.* Nov 16 1996;348(9038):1363-1365.

431. Doubts cast on 1959 AIDS claim. *Science.* Apr 7 1995;268(5207):35.

432. Froland SS, Jenum P, Lindboe CF, Wefring KW, Linnestad PJ, Bohmer T. HIV-1 infection in Norwegian family before 1970. *Lancet.* Jun 11 1988;1(8598):1344-1345.

433. Hooper E. Sailors and star-bursts, and the arrival of HIV. *BMJ.* Dec 20-27 1997;315(7123):1689-1691.

434. CDC. The safety of hepatitis B virus vaccine. *MMWR Morb Mortal Wkly Rep.* Mar 18 1983;32(10):134-136.

435. CDC. Unexplained immunodeficiency and opportunistic infections in infants--New York, New Jersey, California. *MMWR Morb Mortal Wkly Rep.* Dec 17 1982;31(49):665-667.

436. Ammann AJ, Cowan MJ, Wara DW, et al. Acquired immunodeficiency in an infant: possible transmission by means of blood products. *Lancet.* Apr 30 1983;1(8331):956-958.

437. Rubinstein A. Acquired immunodeficiency syndrome in infants. *Am J Dis Child.* Sep 1983;137(9):825-827.

438. Shannon KM, Ammann AJ. Acquired immune deficiency syndrome in childhood. *J Pediatr.* Feb 1985;106(2):332-342.

439. Anderson RM, Medley GF. Epidemiology of HIV infection and AIDS: incubation and infectious periods, survival and vertical transmission. *AIDS.* 1988;2 Suppl 1:S57-63.

440. CDC. A cluster of Kaposi's sarcoma and Pneumocystis carinii pneumonia among homosexual male residents of Los Angeles and Orange Counties, California. *MMWR Morb Mortal Wkly Rep.* Jun 18 1982;31(23):305-307.

441. Auerbach DM, Darrow WW, Jaffe HW, Curran JW. Cluster of cases of the acquired immune deficiency syndrome. Patients linked by sexual contact. *Am J Med.* Mar 1984;76(3):487-492.

442. Isaksson B, Albert J, Chiodi F, Furucrona A, Krook A, Putkonen P. AIDS two months after primary human immunodeficiency virus infection. *J Infect Dis.* Oct 1988;158(4):866-868.

443. Vittecoq D, Autran B, Bourstyn E, Chermann JC. Lymphadenopathy syndrome and seroconversion two months after single use of needle shared with an AIDS patient. *Lancet.* May 31 1986;1(8492):1280.

444. Fincher RM, de Silva M, Lobel S, Spencer M. AIDS-related complex in a heterosexual man seven weeks after a transfusion. *N Engl J Med.* Nov 7 1985;313(19):1226-1227.

445. Pedersen C, Nielsen JO, Dickmeis E, Jordal R. Early progression to AIDS following primary HIV infection. *AIDS.* Jan 1989;3(1):45-47.

446. Jaffe HW, Hardy AM, Morgan WM, Darrow WW. The acquired immunodeficiency syndrome in gay men. *Ann Intern Med.* Nov 1985;103(5):662-664.

447. Curran JW. The epidemiology and prevention of the acquired immunodeficiency syndrome. *Ann Intern Med.* Nov 1985;103(5):657-662.

448. CDC. HIV/AIDS Surveillance Report 1991. January 1992;8(2).

449. Jonassen TO, Stene-Johansen K, Berg ES, et al. Sequence analysis of HIV-1 group O from Norwegian patients infected in the 1960s. *Virology.* Apr 28 1997;231(1):43-47.

450. Getchell JP, Hicks DR, Svinivasan A, et al. Human immunodeficiency virus isolated from a serum sample collected in 1976 in Central Africa. *J Infect Dis.* Nov 1987;156(5):833-837.

451. Zhu T, Korber BT, Nahmias AJ, Hooper E, Sharp PM, Ho DD. An African HIV-1 sequence from 1959 and implications for the origin of the epidemic. *Nature.* Feb 5 1998;391(6667):594-597.

452. Worobey M, Gemmel M, Teuwen DE, et al. Direct evidence of extensive diversity of HIV-1 in Kinshasa by 1960. *Nature.* Oct 2 2008;455(7213):661-664.

453. Katner HP, Pankey GA. Evidence for a Euro-American origin of human immunodeficiency virus (HIV). *J Natl Med Assoc.* Oct 1987;79(10):1068-1072.

454. Biggar RJ, Nasca PC, Burnett WS. AIDS-related Kaposi's sarcoma in New York City in 1977. *N Engl J Med.* Jan 28 1988;318(4):252.

455. (APIIEG) APIiIEGI. Consensus Recommendations for the use of Immunoglobulin Replacement Therapy in Immune Deficiency. 2nd Edition, July 2009 ed; 2009: http://www.apiieg.org/files/1/APIIEG%20Consensus%20Recommendations%20Edition%201%20June%202008.pdf. Accessed May 17, 2009.

456. Notarangelo LD. Primary immunodeficiencies. *J Allergy Clin Immunol.* Feb 2010;125(2 Suppl 2):S182-194.

457. Tarzi MD, Grigoriadou S, Carr SB, Kuitert LM, Longhurst HJ. Clinical immunology review series: An approach to the management of pulmonary disease in primary antibody deficiency. *Clin Exp Immunol.* Feb 2009;155(2):147-155.

458. Cunningham-Rundles C, Bodian C. Common variable immunodeficiency: clinical and immunological features of 248 patients. *Clin Immunol.* Jul 1999;92(1):34-48.

459. Singer C, Armstrong D, Rosen PP, Schottenfeld D. Pneumocystis carinii pneumonia: a cluster of eleven cases. *Ann Intern Med.* Jun 1975;82(6):772-777.

460. Walzer PD, Perl DP, Krogstad DJ, Rawson PG, Schultz MG. Pneumocystis carinii pneumonia in the United States: epidemiologic, diagnostic, and clinical features. *Natl Cancer Inst Monogr.* Oct 1976;43:55-63.

461. Sepkowitz KA, Brown AE, Telzak EE, Gottlieb S, Armstrong D. Pneumocystis carinii pneumonia among patients without AIDS at a cancer hospital. *JAMA.* Feb 12 1992;267(6):832-837.

462. McClure HM, Keeling ME, Custer RP, Marshak RR, Abt DA, Ferrer JF. Erythroleukemia in two infant chimpanzees fed milk from cows naturally infected with the bovine C-type virus. *Cancer Res.* Oct 1974;34(10):2745-2757.

463. Johnson ES. Mortality from non-malignant diseases among women in the meat industry. *Br J Ind Med.* Jan 1987;44(1):60-63.

464. Johnson ES. Mortality among nonwhite men in the meat industry. *J Occup Med.* Mar 1989;31(3):270-272.

465. Johnson ES, Fischman HR, Matanoski GM, Diamond E. Occurrence of cancer in women in the meat industry. *Br J Ind Med.* Sep 1986;43(9):597-604.

466. Bethwaite P, McLean D, Kennedy J, Pearce N. Adult-onset acute leukemia and employment in the meat industry: a New Zealand case-control study. *Cancer Causes Control.* Sep 2001;12(7):635-643.

467. Metayer C, Johnson ES, Rice JC. Nested case-control study of tumors of the hemopoietic and lymphatic systems among workers in the meat industry. *Am J Epidemiol.* Apr 15 1998;147(8):727-738.

468. Parratt D. Nutrition and immunity. *Proc Nutr Soc.* May 1980;39(2):133-140.

469. Higgs JM. Chronic mucocutaneous candidiasis: iron deficiency and the effects of iron therapy. *Proc R Soc Med.* Aug 1973;66(8):802-804.

470. Gracey M, Stone DE, Suharjono, Sunoto. Isolation of Candida species from the gastrointestinal tract in malnourished children. *Am J Clin Nutr.* Apr 1974;27(4):345-349.

471. Paillaud E, Merlier I, Dupeyron C, Scherman E, Poupon J, Bories PN. Oral candidiasis and nutritional deficiencies in elderly hospitalised patients. *Br J Nutr.* Nov 2004;92(5):861-867.

472. Smythe PM, Brereton-Stiles GG, Grace HJ, et al. Thymolymphatic deficiency and depression of cell-mediated immunity in protein-calorie malnutrition. *Lancet.* Oct 30 1971;2(7731):939-943.

473. Gross RL, Newberne PM. Role of nutrition in immunologic function. *Physiol Rev.* Jan 1980;60(1):188-302.

474. Olumide YM. Nutritional dermatoses in Nigeria. *Int J Dermatol.* Jan 1995;34(1):11-16.

475. Drake TE. Thymolymphatic deficiency and depression of cell-mediated immunity in protein-calorie malnutrition. *Lancet.* Dec 11 1971;2(7737):1322-1323.
476. Woodruff JF. Thymolymphatic deficiency and depression of cell-mediated immunity in protein-calorie malnutrition. *Lancet.* Jan 8 1972;1(7741):92-93.
477. Johnson NM. Pneumonia in the acquired immune deficiency syndrome. *Br Med J (Clin Res Ed).* May 4 1985;290(6478):1299-1301.
478. Fukasawa M, Miura T, Hasegawa A, et al. Sequence of simian immunodeficiency virus from African green monkey, a new member of the HIV/SIV group. *Nature.* Jun 2 1988;333(6172):457-461.
479. Kamradt T, Niese D, Vogel F. Slim disease (AIDS). *Lancet.* Dec 21-28 1985;2(8469-70):1425.
480. Tauris P, Black FT. Heterosexuals importing HIV from Africa. *Lancet.* Feb 7 1987;1(8528):325.
481. Biggar RJ. The AIDS problem in Africa. *Lancet.* Jan 11 1986;1(8472):79-83.
482. Kibedi W. AIDS: an African viewpoint. *Dev Forum.* Mar 1987;15(2):1, 6.
483. CIA Factbook Haiti. https://www.cia.gov/library/publications/the-world-factbook/geos/ha.html. Accessed November 7, 2010.
484. CIA Factbook - Zaire (Democratic Republic of the Congo). https://www.cia.gov/library/publications/the-world-factbook/geos/cg.html. Accessed November 7, 2010.
485. Szmuness W, Much I, Prince AM, et al. On the role of sexual behavior in the spread of hepatitis B infection. *Ann Intern Med.* Oct 1975;83(4):489-495.
486. Szmuness W, Dienstag JL, Purcell RH, Harley EJ, Stevens CE, Wong DC. Distribution of antibody to hepatitis a antigen in urban adult populations. *N Engl J Med.* Sep 30 1976;295(14):755-759.
487. Koblin BA, Morrison JM, Taylor PE, Stoneburner RL, Stevens CE. Mortality trends in a cohort of homosexual men in New York City, 1978-1988. *Am J Epidemiol.* Sep 15 1992;136(6):646-656.
488. Altman KA. Rare Cancer seen in 41 Homosexuals. *New York Times*, July 3, 1981.
489. Havens PL. Postexposure prophylaxis in children and adolescents for nonoccupational exposure to human immunodeficiency virus. *Pediatrics.* Jun 2003;111(6 Pt 1):1475-1489.
490. Pinkerton SD, Martin JN, Roland ME, Katz MH, Coates TJ, Kahn JO. Cost-effectiveness of postexposure prophylaxis after sexual or injection-drug exposure to human immunodeficiency virus. *Arch Intern Med.* Jan 12 2004;164(1):46-54.
491. Mbubaegbu C. AIDS in Africa. Inflationary statistics distort the truth. *BMJ.* Jun 19 1993;306(6893):1691.
492. Garry RF, Witte MH, Gottlieb AA, et al. Documentation of an AIDS virus infection in the United States in 1968. *JAMA.* Oct 14 1988;260(14):2085-2087.
493. Dechazal L, Goussard B, Salaun JJ, Bernard J, Zagury D. [Randomization of a population for a clinical trial of immunization against the human immunodeficiency virus (HIV)]. *Med Trop (Mars).* Oct-Dec 1988;48(4):413-416.
494. Nemeth A, Bygdeman S, Sandstrom E, Biberfeld G. Early case of acquired immunodeficiency syndrome in a child from Zaire. *Sex Transm Dis.* Apr-Jun 1986;13(2):111-113.
495. Vandepitte J, Verwilghen R, Zachee P. AIDS and cryptococcosis (Zaire, 1977). *Lancet.* Apr 23 1983;1(8330):925-926.
496. Vangroenweghe D, Afrique. Sese, Bruxelles BEE, c2000. *Sida et sexualité en Afrique EPO;* 2000.

497. Morvan J, Carteron B, Laroche R, Bouillet E, Teyssou R, Blanchard de Vaucouleurs F. [A sero-epidemiologic survey of HIV infection in Burundi between 1980 and 1981]. *Bull Soc Pathol Exot Filiales.* Jan 1989;82(1):130-140.

498. Katlama C, Leport C, Matheron S, et al. Acquired immunodeficiency syndrome (AIDS) in Africans. *Ann Soc Belg Med Trop.* 1984;64(4):379-389.

Appendix — The Earliest Purported Cases of AIDS

The following table was adapted from *Vangroenweghe et al.* (2000) with the addition of two other early purported AIDS cases. The pages following the table contain case-by-case descriptions of the patients' clinical profiles, when available, and commentary by the Author on each case.

Collectively, many of these patients exhibited conditions indicative of impaired cell-mediated immunity. It is likely that some or all of these patients had some form of cell-mediated immunity, but not necessarily HIV infection. (See sub-section *"Other Forms of Immunodeficiency"* regarding other types of immunodeficiency in the chapter *"Outliers – The Exception as the Rule".*)

Table 20 – Earliest Purported African AIDS Cases, 1959 – 1983

Time Period	No. Cases	Location	Methodology	Sub-Type of Extracted Virus
1959	1	Manchester, England	Clinical + Serum	
1959	1	Léopoldville, Belgian Congo	Serum	HIV-I Group M
1960	1	Léopoldville, Belgian Congo	Clinical + Serum	
1961-I962	1	Norwegian infected at Douala, Cameroon)	Serum + Clinical	HIV-I Group O
1962-1976	7	Zaire and Burundi	Serum + Clinical (4) Clinical (3)	
1966-1969	2	wife and child of the Norwegian	Serum + Clinical	HIV-I Group O
1969	1	St Louis, Missouri	Serum + Clinical	
1970	2	Leopoldville	Serum	
1972 - 1977	1	Danish medical doctor infected in Zaire)	Serum + Clinical	
1974	1	Child in Kinshasa (perinatal)	Serum + Clinical	
1975 +	---	Kinshasa: emergence of Kaposi	Serum + Clinical	
1976	5	Yambuku (NW Congo)	Serum	
1977 - 1978	1	Zairean woman treated in Belgium	Clinical	
1978	1	Greek fisherman of Lake Tanganyika	Serum + Clinical	
1978	1	Child of Zaire an parents treated in Sweden	Serum + Clinical	
1979	---	Epidemic of AIDS in Zaire		
1980	3%	Zairean mothers in Leopoldville	Serum	
1980-1981	29	Burundi (4.4% of examined sera)	Serum	
1982-1983	9	Zaireans treated in Paris	Clinical	
	4	Congolese (Brazzaville)	Clinical	
	1	Mali	Clinical	

Source: Vangroenweghe D. The earliest cases of human immunodeficiency virus type 1 group M in Congo-Kinshasa, Rwanda and Burundi and the origin of acquired immune deficiency syndrome. Philos Trans R Soc Lond B Biol Sci. Jun 29 2001;356(1410):923-925.

In Table 19, above, the Methodology column lists "Clinical" and/or "Serum." Clinical designates that the retrospective diagnosis was made on the basis of a clinical report, i.e., records of the patients' disease condition(s). Serum indicates that: (1) a stored blood or tissue sample was tested for the HIV antibody; and/or (2) HIV was purportedly derived from a stored blood or tissue.

The column "Sub-Type of Extracted Virus" indicates that HIV and/or its genetic fragments purported was isolated from a blood or tissue sample. Of the three HIV groups, Group M (major) is the overwhelmingly predominate viral group; it is responsible for HIV infection throughout the United States, the Caribbean, the Americas, Europe, and Central Africa (meaning Zaire, primarily). Group O (outlier) is self-descriptive; and Group N (new), the group seemingly established around 1999 when it contained viruses documented in only two Cameroon individuals at that time.[136]

Time Period:	1960 – 1962
Number:	1
Location:	English Sailor, Manchester England
Methodology:	Clinical Report + Serum Analysis

Description:

In one prominent and widely publicized case, a seaman from Manchester, England reportedly had AIDS in 1959. He had been diagnosed with Pneumocystis carinii infection and the case had stumped his doctors. In 1983, an attending physician wrote a letter to the scientific journal Lancet entitled: "AIDS in 1959?" He described a clinical profile matching the AIDS prodrome and the autopsy finding of PCP. The patient had been in the Navy between 1995–1957 and "had traveled abroad (the presumption being Africa?)."[424]

In 1990, scientists reportedly extracted HIV from the seamen's preserved tissue. In the end, this "discovery" of HIV from 1959 was actually another incident of laboratory contamination.

It was also later clarified that although the English sailor in the Royal Navy, but "was stationed exclusively in England, except for a brief voyage to Gibraltar in 1957."[429]

As usual, the retraction gained far less notoriety than the initial false finding. Today, the belief in the validity of this disproved AIDS case stubbornly persists among the lay and professional communities alike.

References:

Williams G, Stretton TB, Leonard JC. AIDS in 1959? Lancet. Nov 12 1983;2(8359):1136 • Williams G, Stretton TB, Leonard JC. Cytomegalic inclusion disease and Pneumocystis carinii infection in an adult. Lancet. Oct 29 1960;2(7157):951-955 • Corbitt G, Bailey AS, Williams G. HIV infection in Manchester, 1959. Lancet. Jul 7 1990;336(8706):51 • Corbitt G, Bailey AS. AIDS in Manchester, 1959? Lancet. Apr 22 1995;345(8956):1058. Bailey AS, Corbitt G. Was HIV present in 1959? Lancet. Jan 20 1996;347(8995):189 • Zhu T, Ho DD. Was HIV present in 1959? Nature. Apr 6 1995;374(6522):503-504 • Hooper E, Hamilton WD. 1959 Manchester case of syndrome resembling AIDS. Lancet. Nov 16 1996;348(9038):1363-1365 • Doubts cast on 1959 AIDS claim. Science. Apr 7 1995;268(5207):35 [424-431]

Comment: This patient apparently had some form of cell-mediated immunodeficiency, but it was unlikely to have been HIV infection.

Time Period:	1959
Number:	1
Location:	Léopoldville, Belgian Congo*
Methodology:	Serum Analysis

Description:

HIV purportedly isolated from blood sample obtained Léopoldville in 1959. Briefly discussed in the sub-section *"Prominent Outliers"* in the chapter *"Outliers – The Exception as the Rule."*

A total of 1,213 plasma samples obtained in Léopoldville between 1959 and 1982 were evaluated by immunoassay. The immunoassay found 21 seropositive samples, but only 1 was confirmed as reactive with HIV-1 by immunofluorescence, western blotting and radioimmunoprecipitation methods.

This positive plasma sample (L70) was obtained in early 1959 from an adult Bantu male, with a sickle-cell trait and a glucose-6-phosphate-dehydrogenase deficiency, living in Leopoldville, Belgian Congo (now Kinshasa, Democratic Republic of Congo).

Sample L70 was analyzed for the possible presence of RNA sequences common to HIV. Chemically, these researchers cast a wide net with the "probes" they used:

> "Because of the limited amount of plasma available from this sample and uncertainty about its condition, efforts were made to increase the likelihood of recovering HIV-1 sequences by RT-PCR (reverse transcription followed by polymerase chain reaction). Multiple primers were used in a single RT reaction, and all synthesized complementary DNAs were amplified by PCR using primers designed to amplify HIV-1 sequences from all known subtypes. However, attempts to amplify HIV-1 fragments of .300 base pairs (bp) were unsuccessful, probably because of a low level of intact viral RNA in this old sample. However, after numerous attempts, four shorter sequences were obtained."

> Multiple phylogenetic analyses not only authenticate this case as the oldest known HIV-1 infection, but also place its viral sequence near the ancestral node of subtypes B and D in the major group, indicating that these HIV-1 subtypes, and perhaps all major group viruses, may have evolved from a single introduction into the African population not long before 1959."

Reference: Zhu T, Korber BT, Nahmias AJ, Hooper E, Sharp PM, Ho DD. An African HIV-1 sequence from 1959 and implications for the origin of the epidemic. Nature. Feb 5 1998;391(6667):594-597 [451]

Comment: These investigators use a common technique to fish for any fragments of RNA that contain sequences common to HIV. These fragments are then "amplified," meaning the individual fragments were copied and assembled together into a strand of sufficient length such that strand can undergo more detailed investigations into its genetic structure. After PCR amplification of these fragments, the PCR products are compared to a library of genetic sequences associated with HIV.

There are several possible interpretations of these findings. The investigators are basically interpreting the findings as evidence that these fragments emanate from progenitor of HIV. They do so, of course, in light of the theory that HIV and AIDS originated in Africa, a comparative genetic library formed on the same premise.

If this is a legitimate sample, that is, the viral fragments in the sample were actually derived from the patient samples (and not a laboratory contaminant), then these viral fragments may represent a historical retrovirus that had some of the same conserved elements of HIV, i.e., they share some degree of genetic homology with HIV. Homology is relative, for example, humans, chimpanzees, gorillas, monkeys, dolphins and whales are share a skeletal homology, indicating common ancestry (dolphins and whales have vestigial hands and feet).

However, the timing and placement of this sample makes the presence of HIV unlikely.

Conversely, if these viral elements were actually fragments of HIV, then it seems far more likely that these viral fragments arose from laboratory contamination than from one isolated individual in 1959.

---

*now Kinshasa, Democratic Republic of the Congo

Time Period:	1960
Number:	1
Location:	Léopoldville, Belgian Congo
Methodology:	Clinical Report + Serum Analysis

Description:

The investigative team obtained 813 paraffin-embedded histopathological blocks, meaning tissue samples preserved in paraffin ("wax") from the 1958–1962 archives of the Department of Anatomy and Pathology at the University of Kinshasa.

They used 14 primer sets designed to anneal to highly conserved regions of the *gag*, *pol* and *env* genes of HIV-1 group M and to amplify very short fragments likely to be present even in ancient and/or degraded specimens.

They amplify and characterize viral sequences derived from a Bouin's-fixed paraffin-embedded lymph node biopsy specimen obtained in 1960 from an adult female, and conduct the first comparative evolutionary genetic study of early pre-AIDS epidemic HIV-1 group M viruses.

They conclude the viral sequences belong to HIV-1 Group M (Main), subtype A.

Reference: Worobey M, Gemmel M, Teuwen DE, et al. Direct evidence of extensive diversity of HIV-1 in Kinshasa by 1960. Nature. Oct 2 2008;455(7213):661-664. [452]

Comment:

This study was a fishing expedition among whatever samples were available for prospecting. They searched with the expectation of finding a precursor to HIV and found one. This was apparently a blind sample without any patient history attached for the patient population and/or the female in question except, presumably, they were pathological samples for diseases patients.

The investigators derive and amplify viral sequences and compare them with a library of sequences while performing convoluted analyses entailing complex models, comparisons, and assumptions.

In the end, they conclude these viral fragments derived from this fixed sample were the same strain of HIV common to the HIV/AIDS epidemic in the U.S. and Europe.

---

*now Kinshasa, Democratic Republic of the Congo

Time Period:	1960 – 1962
Number:	1
Location:	Norwegian
Methodology:	Clinical Report + Serum Analysis

Description:

This Norwegian is the father of the Norwegian family described in sub-section *"Prominent Outliers"* in the chapter *"Outliers – The Exception as the Rule."* Upon serum analysis, the virus or viral fragment purportedly isolated from this patient belonged to HIV-1, Group M (major), the most common HIV strain found in among HIV/AIDS patients in the United States and Europe.

Reference: Froland SS, Jenum P, Lindboe CF, Wefring KW, Linnestad PJ, Bohmer T. HIV-1 infection in Norwegian family before 1970. Lancet. Jun 11 1988;1(8598):1344-1345 [432]

Comment: See description in sub-section *"Prominent Outliers"* in the chapter *"Outliers – The Exception as the Rule."*

Time Period:	1969
Number:	1
Location:	St Louis, Missouri
Methodology:	Clinical Report + Serum Analysis

Description:

A 15-year-old black male who was admitted to St Louis City Hospital in 1968 for extensive lymphedema ("swelling") of the genitalia and lower extremities. *Chlamydial* organisms were widely disseminated and isolated from numerous body fluids and organs. Over a 16-month clinical course his condition progressively deteriorated, and at autopsy there was widespread Kaposi's sarcoma of the aggressive, disseminated type. Recently performed Western blot and antigen capture assays on serum and autopsy tissue specimens frozen since 1969 have disclosed that this sexually active teenager was infected with a virus closely related or identical to human immunodeficiency virus type 1. The clinical and immunologic findings together suggest that an immunosuppressive retrovirus existed in the United States before the late 1970s.

Reference: Garry RF, Witte MH, Gottlieb AA, et al. Documentation of an AIDS virus infection in the United States in 1968. JAMA. Oct 14 1988;260(14):2085-2087 [492]

Comment: *Chlamydia* is a bacterium, so not a typical presenting opportunistic infection of HIV infection, but the aggressive disseminated KS found upon autopsy indicates an impairment of cell-mediated immunity, but apparently a selective one.

The patient also had seropositive reactions for herpes simplex virus, cytomegalovirus, and Epstein-Barr virus, although none of these infections had reactivated. Perhaps this patient's body and/or severe morbidity confounded all the viral assays.

The patient had no history of travel outside of the Midwest, intravenous drug abuse, or blood transfusion.

This seems a rare case of cell-mediated immunity consequent to an unknown etiology; the presence of HIV seems unlikely due to any recognizable transmission vector that would put this individual at specific risk in absence of a greater threshold population.

Time Period:	1962-1976
Number:	7
Location:	Zaire + Burundi
Methodology:	Clinical Report + Serum Analysis (4)
	Clinical Report (3)

Description:

Case 1 — an ill 50-year-old black woman died of disseminated Kaposi's sarcoma (KS).

She had been ill many other respects: fever, purulent stomatitis, enlarged cervical lymph nodes, respiratory distress, cachexia, pitting edema of the legs; purulent gingivitis, ulcerative stomatitis, and severe halitosis. Culture of the purulent exudates from mouth and neck excluded a mycotic or a tuberculosis infection. They yielded *Bacteroides*, anaerobic, and beta-hemolytic streptococci and *Fusobacterium* species.

"Despite the lack of serological and immunological evidence, the clinical features appear at least as very suggestive of AIDS. Indeed, Case 1 fulfils the requirements of the CDC AIDS case definition."

Comment, Case 1 — the patient was obviously in the terminal stage of some illness; and was infected by several colonies of bacteria; not necessarily an indication of early HIV infection; but such infections are more likely in the terminal stage of HIV infection due to an overall degradation of bodily functions at the approach of death.

Owing the presence of fatal, disseminated KS, this woman evidently had some form of defect in her cell-mediated immunity; however, the timing and placement of this case makes the presence of HIV unlikely.

These authors also report: "There was no outbreak of AIDS-like cases in the Kinshasa area during the period 1957-1966 and aggressive KS in young adults was exceedingly rare." Thereby, rendering this purported HIV/AIDS cases as a singular disparate case of a purportedly communicable disease lacking a cluster. More likely, it is an incident of severe cell-mediated immunity due to and uncharacterized primary or secondary etiology.

Case 2 — a white male, born in Belgium; had been living in Zaire and Burundi as a building contractor 1971-1976; married a Rwandan woman in 1973 (Case 3); wife declared her late husband was not associated with any of the risk factors for AIDS: male homosexuality, intravenous drug abuse or previous blood transfusions.

In early 1976, the first symptoms suggestive of HIV infection appeared: persistent cervical lymph node enlargement, facial dermatitis, and a herpes on the thigh. The patient had reduced response to a control mitogen. A mitogen is a substance used to test immunological allergic response; a reduced response to common mitogens is common during HIV infection.

Biopsy of lymph tissue performed in 1986 revealed follicular hyperplasia characterized as AIDS-like. The patient died of cerebral *Toxoplasma gondii* infection and pneumonia in 1981 after exhibiting mental derangement and fever (*T .gondii* antibodies had been found in 1976).

Comment, Case 2 — *T. gondii* is protozoal parasite, found in domestic and wild animals, such as cats. Humans may catch from *T. gondii* droppings of cats and undercooked meat, especially mutton. In AIDS patients, tendency to infect tissues of central nervous system (brain and nerves). *T. gondii* infection may also cause pneumonia and hepatitis inflammation/dysfunction of the liver). Many minor, non-life-threatening outbreaks occur in day-care centers. In AIDS patients, can be a major cause of mortality.

T. gondii infection (toxoplasmosis) is considered an AIDS-defining condition. IN PLWH, toxoplasmosis frequently infects the brain or lungs. However; toxoplasmosis can occur in absence of HIV infection. Toxoplasmosis also occurs in patients after organ transplantation wherein they receive immuno-suppressive drug therapy to prevent organ rejection.

Overall, this patient had a toxoplasmosis which can be AIDS-defining infection; and had infections that typified HIV infection. However, the timing and placement of this case makes the presence of HIV unlikely. Therefore, it is likely that this patient had some other form of impaired cellular-immunity. Fatal toxoplasmosis prior to the advent of AIDS was rare, but not unknown.

Case 3 — a Rwandan woman; the widow of Case 2 whom she married at 17 years of age; mother of 3 children born.

She presented with hypergammaglobulinemia (excess of antibodies, common to HIV infection but perhaps more common in infants); and later with persistent generalized lymphadenopathy. She tested seropositive in August 1984 and December 1985. HIV was isolated by culture of peripheral lymphocytes in February 1986.

The patient denied drug addiction and had never been transfused. She had no risk factor other than heterosexual exposure during her married life time but admitted rare sexual contacts with a heterosexual seronegative white partner since her widowhood. The T4 count was lowered with a range of 150 to 350 μL and the fluctuating T4/T8 ratio ranged from 1.1 to 0.67. Her response to mitogens was normal.

All children seronegative for HIV in 1986.

Comment, Case 3 — a rather confounding case; particularly given the proximity of her partner. The patient purportedly had HIV infection for a minimum of 5 years without manifesting opportunistic diseases. Her immunological profile was poor but not extreme, and fluctuating into the lower norms of normal. Her lymphocyte proliferative response to mitogens was "normal" which is not typical for HIV infection.

Technically, the hypergammaglobulinemia and her disturbed immunological profile could be consequent to HIV infection, or to some yet unidentified condition. Both she and her husband's stored tissues tested seropositive. In the United States at this time, 1 in 100 people tested positive (where as only 0.035 % were confirmed by follow-up testing). Assays in themselves are not necessarily definitive, and some populations with high prevalences of serious illnesses seem to test positive at inordinate levels.

Blatantly, the Author is incredulous of HIV being isolated from this patient's samples. Rather, the Author suspects this HIV is a laboratory artifact, i.e. a contaminant. It was not uncommon at that time. In the 1988 *Nature* discussion of the laboratory contamination that spawned the fallacious theory that HIV originated in African green monkeys (and subsequent African primates; see the chapter "*The Contaminated Monkey Theory*"). Notably, the author of the *Nature* article cited in this chapter stated that he knew of 5 other instances of contamination in the United States and Europe wherein cell cultures by other cell cultures under the same biological safety hoods.

The timing and placement of this case makes the presence of HIV unlikely.

Case 4 — a white Belgian male residing in Shaba (former Katanga, Zaire); married to a Belgian woman; they had one child. In 1975, he presented with rhinitis and hearing problems; and lymphatic non-malignant hyperplasia, chorioretinitis (a disease common to HIV infection); a feverish rash; CNS symptoms, including epileptic seizures, left upper limb palsy and progressive coma. He died in late 1977. At autopsy, numerous *T. gondii* abscesses were discovered in both cerebral hemispheres

Comment, Case 4 — As in Case 2, above, this patient died of cerebral toxoplasmosis; a disease that can occur without HIV infection. The course of the disease was rapid; but absent of any other AIDS-defining opportunistic infections.

The timing and placement of this case makes the presence of HIV unlikely.

Case 5 — a heterosexual white male Belgian was discovered as seropositive during routine voluntary blood donation in September 1985. He had divorced 5 years earlier; denied homosexuality, bisexuality, present or past intravenous drug addiction; and blood transfusion. The patient had repeated heterosexual intercourse with Zairian prostitutes when he participated in a 2-year voluntary service overseas (Zaire), from 1976 – 1978.

However, 3 surgical patients, not at risk for AIDS, who had received whole blood (1 case) or fresh plasma (2 cases) from this donor were found to be asymptomatic HIV antibody carriers throughout a 6-month period following transfusion. Later (October 1986) case 5 remained asymptomatic, while the T4 count and the T4/T8, ratio had dropped from 567 to 333/pl and from 0.58 to 0.25 respectively.

The investigators attempted and failed to isolate HIV from the patient's lymphocytes. Attempts to culture viruses can be hit-or-miss affairs. It is possible that a virus is present in a patient but one can repeatedly fail to culture it.

Comment, Case 5 — another confounding case. To have theoretically contracted HIV in Africa during 1976 – 1978 seems unlikely, as it is unlikely HIV would have been present in Central Africa at that time. He was asymptomatic for 7 – 9 years after feasible infection, and his immunological profile was odd with low but sufficient T4 cells and a low T4/T8 ratio. His lymphocyte proliferative response to mitogens was "unaltered," which is not typical for HIV infection.

A rather indeterminate case.

Whatever antibodies or substances this man carried in his blood were evidently transmitted to the recipients of his blood donation. HIV antibodies has been passively transmitted by blood-derived products (meaning only the anti-HIV antibodies where transmitted to the recipient, but HIV itself was not present). Perhaps this patient had some viral or retroviral or some other infection obtained in his time in Africa that generated antibodies or other substances that fooled the HIV assay.

If he had manifested disease, one might question the validity of his statement that he was exclusively heterosexual, because HIV was definitely among the Belgium population by 1985. Otherwise, it appears a random false alarm. In the United States at this time, 1 in 100 people tested positive (where as only 0.035 % were confirmed by follow-up testing).

Case 6 — a black Zairian woman, married to a Belgian colonial officer (Case 7); mother of 6 children; resident of Belgium since 1968; never returned to Central Africa. During numerous interviews the patient denied previous transfusions, drug abuse and extramarital sexual intercourses. The sincerity of the patient's statements was ascertained by her family physician.

In December 1981, at the age of 50, she developed dental abscess; developed a febrile illness with general malaise; enlargement of the cervical and subclavicular lymph nodes; a febrile aseptic meningitis; and elevated levels of white blood cells, and cytomegalovirus (CMV) infection. The patient recovered but had persistent generalized lymphadenopathy for at least 3 years. In 1982, cervical biopsy showed hyperplasia; later identified as the follicular type characteristic of HIV infection and she had hypergammaglobulinemia. In 1984, her T4 cell count as 213 µL and her T4/T8 ratio 1.03. In 1985, she was asymptomatic and most of the enlarged lymph nodes had disappeared.

During the years 1985 and 1986, her health deteriorated. In mid-1986, the patient suffered from *Candida albicans* oesophagitis and from diarrhoea due to *Isospora belli* and *Campylobacter jejuni*. The first two are AIDS-defining conditions, the later a common human bacterial infection. Her T4 cell count had decreased to 15 µL and the T4/T8 ratio 0.023. Her lymphoproliferative response to the mitogens was "reduced."

In 1986 her 6 children, 13-34 years old, were found to be in a good state of health and HIV seronegative.

Comment, Case 6 — a rather confounding case of again a married couple having suspected immunodeficiency due to HIV infection. This patient apparently had impaired cellular immunity that gave rise to AIDS-defining conditions. It seems odd that the CMV infection did not reactivate in the presence of this evident cellular immunity, but such is the nature of a syndrome: not all possible components need be present in any one patient.

Although this woman resided in Europe during the time that HIV had been introduced to Europe from the United States, her demographic profile does not place her amongst the high-risk categories, so it is unlikely she personally contracted HIV in Europe. If she or husband contracted HIV in Africa, it was prior to 1968, when they came to Europe on her husband's retirement. The timing and placement of this sample makes the presence of HIV unlikely. Also, the incubation period would have been minimally 13 years.

Case 7 — a white Belgian, husband of Case 6; retired after a 22-year career as a colonial officer in Zaire, found to be an asymptomatic HIV antibody carrier at the age of 71. He denied any risk factor for AIDS except past heterosexual exposure. The physical examination remained normal until end of 1986.

However, his T4 cell count was 238 15 µL and the T4/T8 ratio 0.24 at the end of 1986. The immunity parameters remained stable during the period of study (1985-1986), without any symptoms suggesting an impaired cellular immunity. The investigators attempted and failed to isolate HIV from the patient's lymphocytes. Attempts to culture viruses can be hit-or-miss affairs. It is possible that a virus is present in a patient but one can repeatedly fail to culture it.

Comment, Case 7 — another unlikely candidate for HIV infection, despite the seropositive result in the HIV assay. He has a disturbed immunological profile for some unknown reason but has yet to manifest opportunistic infections over the presumed 13 year incubation period.

Reference: Sonnet J, Michaux JL, Zech F, Brucher JM, de Bruyere M, Burtonboy G. Early AIDS cases originating from Zaire and Burundi (1962-1976). Scand J Infect Dis. 1987;19(5):511-517 [423]

Time Period:	1966 – 1969
Number:	2
Location:	Wife + Child of Norwegian
Methodology:	Clinical Report + Serum Analysis

Description:

This woman and child are the wife and daughter of the aforementioned Norwegian man. All three are described in sub-section *"Prominent Outliers"* in the chapter *"Outliers – The Exception as the Rule."*

The strain of HIV-1 purportedly isolated from these patients was Group O (Outlier). The name is self-descriptive.

Reference: Froland SS, Jenum P, Lindboe CF, Wefring KW, Linnestad PJ, Bohmer T. HIV-1 infection in Norwegian family before 1970. Lancet. Jun 11 1988;1(8598):1344-1345.[432]

Comment: See description in sub-section *"Prominent Outliers"* in the chapter *"Outliers – The Exception as the Rule."* .

Time Period:	1970 **X**
Number:	2 **X**
Location:	Léopoldville, Belgian Congo **X**
Methodology:	Serum Analysis **X**

Description:

X - error – these purported AIDS cases were listed by Vangroenweghe et al. (2000). However, Vangroenweghe et al. apparently made an error in citing these cases

Vangroenweghe et al. lists 2 purported AIDS cases in 1970 for this location.

However, the cited reference (Dechazal et al.) does not describe individual cases. Rather it reports a Zairean sero-survey conducted on samples collected between 1986 and 1988.

Dechazal et al. reports that, in Zaire, seropositive prevalence differed from 2.4% (rural population) to 12.5% (urban population) according to the regions. When the group with 2.4% migrated to the area with 12.5% positives, after 8-12 months the number of seropositives in this group rose to 8%, showing an increase of 5.6% within one year.

A very interesting finding. These high HIV seroprevalence rates are no doubt based in inordinate false-positive rates, but are very interesting that something in the urban setting induced an increased prevalence of false-positives outcomes in this population. A tripling of seroprevalence rates due to sexual activity seems unlikely.

This situation suggests some environmental and/or transmissible factor (something universal such as sanitation, food, water, or pest/animal vectors) in the urban Zairean environment can increase false-positive outcomes in a population within a 12-month period. (If transmissible, sexual transmission seems unlikely to effect such a population change in short notice; more likely.

Reference: Dechazal L, Goussard B, Salaun JJ, Bernard J, Zagury D. [Randomization of a population for a clinical trial of immunization against the human immunodeficiency virus (HIV)]. Med Trop (Mars). Oct-Dec 1988;48(4):413-416 [493]

Comment: Outcomes of HIV assays of this generation were highly questionable. See as discussed in various sub-sections of "*HIV Antibody Testing – An Avalanche of False-Positives.*"

---

*now Kinshasa, Democratic Republic of the Congo

Time Period:	1972
Number:	1
Location:	Zaire/Denmark
Methodology:	Clinical Report

Description:

Another prominent case is that of a Danish surgeon, a previously healthy 47-year old woman who died in 1977. She had worked as a surgeon at a primitive, rural hospital in Zaire in 1972–1975. The assumption is that she contracted HIV by being exposed to blood and/or bodily substances during surgery under these conditions. In 1976, she presented with diarrhea, fatigue, wasting, and lymphadenopathy; the conditions resolved but a year later she developed PCP and oral candidiasis.

Reference: Bygbjerg IC. AIDS in a Danish surgeon (Zaire, 1976). Lancet. Apr 23 1983;1(8330):925 [49]

Comment: Very likely, she had some form of impaired cellular immunity, but the timing and placement of this condition makes the presence of HIV unlikely.

Time Period:	1974
Number:	1
Location:	Kinshasa, Zaire
Methodology:	Clinical Report + Serum Analysis

Description:

An 8-year old child born in Zaire to Zairean parents of good socioeconomic status; no hereditary diseases; died in Sweden 1982

At 5 months, the boy had acute viral infection with rash followed by chronic cough, recurrent septicemia, fever, with miliary lung infiltrates, disseminated lymphadenopathy, hepatosplenomegaly, candidiasis, diarrhea, and lethal disturbances of central nervous system. The boy's serum samples from 1981 and 1983 were retrospectively assayed for antibodies to HTLV-III (HIV) and found positive.

"Our case fulfills the criteria of AIDS and has no resemblance to any of the known congenital immunodeficiency syndromes."

Reference: Nemeth A, Bygdeman S, Sandstrom E, Biberfeld G. Early case of acquired immunodeficiency syndrome in a child from Zaire. Sex Transm Dis. Apr-Jun 1986;13(2):111-113 [494]

Comment: A strange case that alternately could likely be a legitimate case of HIV infection, or not. The clinical, immunologic, and laboratory profile (including conditions not described herein) match the profile of HIV infection. However the timing and placement of this sample makes the presence of HIV possible but unlikely.

The physicians conclude the child was infected in Zaire prior to arriving in Sweden in 1978, possibly as early as 1975 in conjunction with an illness (they seem to be inferring the child had a blood transfusion at that time, but don't specifically report one). The child died in September 1982. A rather long incubation period.

Although this case clinically fulfills the concurrent CDC surveillance definition for AIDS, and blood tested retrospectively positive, one has the right to conclude it is a case of HIV infection, except the timing puts it right on the edge of feasibility, but also in the face of some major exceptions.

First, HIV might have entered the human domain by 1978, but first in New York City. Since NYC is an international travel center, is it possible that any visitors or residents at the time could have harbored HIV infection, but the prevalence level was extremely slow. Generally, the AIDS epidemic blossomed first in the U.S., and then migrated over seas, but at least 1 foreign case was concurrent with the first identified blossoming clusters in NYC and Los Angeles — the first AIDS case in Denmark, contracted by a man who had visited the NYC bath houses annually on vacation for the prior ten years. There might have been others.

This child, at 8 years old, seems an unlikely case of perinatal HIV transmission (pregnant mother to birthed child transmission), and the attending physicians agree this form of HIV transmission seems unlikely, citing another report or 29 U.S. infants whose mean incubation period was 4 months (range 1 to 9 months).

Early in the epidemic, the earliest African cases typically occurred in upper class men, their wives, and their children (born with HIV infection). Based on this profile, the general medical consensus developed that HIV was endemic in Africa. However, these incipient AIDS patients traveled between countries at will, and were residents of Europe, or were diagnosed in Europe because they had come seeking medical treatment.

The simplest explanation is that the men in these families were closeted gays. They had wives and children, but had secret lives engaging in homosexual sex in Europe, which is how they contracted HIV infection (not from monkeys). They then transmitted HIV to their wives and/or girlfriends.

In Haiti, heterosexual men contracted HIV by prostituting themselves to gay male tourists from the United States. These heterosexual men then transmitted HIV to their wives and/or girlfriends.

This child's parents match the profile of the typical initial African patients, but the timing of presumed HIV infection of the father would have been very early in the epidemic.

The odds are low that the father of this family was a closeted gay who picked up HIV infection from a well-traveled, social equal in Zaire. Possible but a statistically low chance.

However, the African populations that seemed to typify this behavior were involved primarily in the Francophone (French-speaking) transmission centers and vectors, i.e., the Paris/Brussels Francophone epicenter and the Francophone former French and Belgian colonies. This family in question is Zairean, a Francophone former colony of Belgium, but they came to Sweden seeking medical treatment for their child.

The odds are rather low for a 6-year longevity for a child infected at birth or 5 months. Also, excluding incest by an infected father, or the unspecified possibility of a blood transfusion, no clear or likely route of transmission is apparent.

Time Period:	1975 + ✗
Number:	Emergence of Kaposi's sarcoma ✗
Location:	Kinshasa, Zaire ✗
Methodology:	Clinical Report + Serum Analysis ✗

✗ - error – these purported AIDS cases were listed by Vangroenweghe et al. (2000). However, Vangroenweghe et al. apparently made an error in citing these cases

Description:

The original table reads: "Kinshasa: emergence of Kaposi's sarcoma"

Reference: Piot P, Quinn TC, Taelman H, et al. Acquired immunodeficiency syndrome in a heterosexual population in Zaire. Lancet. Jul 14 1984;2(8394):65-69 [58]

Comment:

This listing of 2 individual AIDS cases in 1970s must stem for a poor reading and/or comprehension of the source article, or the author became confused while listing his references.

Although Piot et al. cited reports of KS among Zairean populations in 1975, a seminal publication that overviews KS prevalence and manifestation across Africa was published in 1962; describing the manifestation and prevalence throughout black and white African populations (Oettle AG. Geographical and racial differences in the frequency of Kaposi's sarcoma as evidence of environmental or genetic causes. Acta Unio Int Contra Cancrum. 1962;18:330-363).

The **CDC** cited Oettle et al. in the first **MMWR** report of a KS outbreak among gay men in July 1981. The CDC cites Oettle in its statement that KS occurs in: "endemic belt across equatorial Africa, where KS commonly affects children and young adults and accounts for up to 9% of all cancers."

So this listing in the table of earliest cases is meaningless. KS was present in the Zairean population in 1975, but KS is endemic to the region. Aggressive cases were rare before the advent of HIV/AIDS, but not non-existent.

Time Period:	1976
Number:	5
Location:	Yambuku (NW Congo)
Methodology:	Serum Analysis

Description:

Serum samples (*n* = 659) collected in the remote Equateur province of Zaire during 1976 were tested for HIV antibodies in 1985.

Five (0.8%) of the samples were seropositive. HIV was purportedly isolated from a blood sample one of these five people.

Follow-up investigations in 1985 revealed that three of the five seropositive persons had died of illnesses suggestive of acquired immunodeficiency syndrome (AIDS), and two remained healthy but seropositive.

> **Subject 1:** a 59-year-old woman; ratio of helper to suppressor T cells (CD4/CD8) was normal
>
> **Subject 2:** a 57-year-old man; ratio of helper to suppressor T cells (CD4/CD8) was abnormally low
>
> **Subject 3:** dead at 38; after a prolonged illness characterized by weight loss, fever, cough, and diarrhea; She had lived in Kinshasa from 1972 to 1976, where she was unmarried and was considered a "free woman." HIV had been isolated from her serum in 1976.
>
> **Subject 4:** dead at 48; she died after a long illness associated with fever, weight loss, skin rash, and oral lesions; wife of Subject 2
>
> **Subject 5:** died of pneumonia and weight loss at the age of 16

With the exception of Subject 3, none of these seropositive persons had traveled outside the region of their respective home villages.

Reference: Nzilambi N, De Cock KM, Forthal DN, et al. The prevalence of infection with human immunodeficiency virus over a 10-year period in rural Zaire. N Engl J Med. Feb 4 1988;318(5):276-279 [309]

Comment:

First, the population under study has a life expectancy of 50 years. Second, all the patients who died exhibited the conditions of "African AIDS" rather than typical early onset symptoms of "classical AIDS," i.e., not the clinical profiles of American or European patients (as described in "African versus Classical AIDS"). Third, given the lack of international travel among this population, and the presumed lack of extensive high traffic from NYC or Paris to these remote locations, it is unlikely that HIV could have arrived in these village as early as 1976.

Finally, the HIV strain derived from a member of this cohort should be genetically compared with all HIV strains previously resident in the analytical laboratory in which the virus was purported derived from this 1976 sample.

Time Period:	1977
Number:	1
Location:	Belgium
Methodology:	Clinical Report

Description:

"A 34-year-old black woman, secretary in an airline company, came from Kinshasa to Belgium in August 1977 seeking medical advice about her 3-month-old daughter who had had oral candidiasis since birth. The woman had 3 healthy children by her first husband. The first child of her second marriage had died at the age of 6 months from a respiratory infection, and the second had died from "septicaemia." Both children had oral thrush. Her third child was seen at our clinic and was found to have a depressed cell-mediated immune response. Treatment with oral miconazole was successful and the immune status of the child was normal when re-tested 2 years later."

While her child was still under investigation, the patient began to complain of fever, fatigue, headache, and pansinusitis. She was first seen as an outpatient on Aug 24, 1977. Iron deficiency and oral candidiasis were diagnosed.

On Sept 19 she was admitted with a fever, rigors, and signs of pulmonary infection, generalized lymphadenopathy, weight loss, and polyarthralgia. Following infections included: oral candidiasis, genital and perianal herpes, and generalized infection with *Cryptococcus neoformans*, repeatedly isolated from CSF, blood, and urine. Blood cultures were successively positive for *Candida albicans, Staphylococcus aureus*, and *enterococci*. Other infections included severe diarrhoea with *Salmonella montevideo* and urinary infections with *Escherichia coli* and *Pseudomonas aeruginosa*.

She died in February, 1978.

"Although the immunological evaluation is incomplete by today's standards this case meets the essential criteria for AIDS. The nature of the immune dysfunction observed in her three children and the genetic or other relationship with the condition in their mother is far from clear, as it was in other infants born from AIDS mothers and observed in the USA."

"A prospective study on cases of cryptococcosis in Kinshasa might unravel some of the mysteries surrounding AIDS."

Reference: Vandepitte J, Verwilghen R, Zachee P. AIDS and cryptococcosis (Zaire, 1977). Lancet. Apr 23 1983;1(8330):925-926 [495]

Comment:

Cryptococcus was rare but not unheard of prior to the advent of AIDS. In the National Library of Medicine, the search term "cryptococcus" gives rise to 1600 articles published prior to 1979. *Cryptococcus* is a fungus, so its infectious presence suggests an underlying immunodeficiency; however, that underlying deficiency need not necessarily induced by HIV infection.

Initially, the patient had iron deficiency which is known to impair cell-mediated immunity such that oral candidiasis develops. Vitamin treatment evidently cleared the problem of this episode.

Later, the patient had both infections considered AIDS-defining and bacterial infections, the latter not indicative of HIV infection as early presentations in themselves, but possible and likely during the terminal stages of infection. Whatever the etiology of this woman's disease, she was in its terminal stages.

However, the timing and placement of this sample makes the presence of HIV unlikely. Given what appears to be a cluster of a mother and two offspring, out of in infectious consideration, the possibility of exposure to some immunotoxin might be investigated; or, possibly, some other infectious agent and/or genetic component might be involved.

Time Period:	1978
Number:	1
Location:	Lake Tanganyika, Central Africa
Methodology:	Clinical Report + Serum Analysis

Description:

The description in the original table was "Greek fisherman of Lake Tanganyika."

Per Wikipedia, Lake Tanganyika is divided among four countries: Burundi, Democratic Republic of the Congo (DRC), Tanzania and Zambia,

Reference: *The Coming Plague: Newly Emerging Diseases in a World Out of Balance* by Laurie Garrett. Penguin. 1995.

Comment:

The cited reference book, *The Coming Plague,* was not available to the Author, partially due to an oversight on the part of the Author, and also the low priority of investigating this particular unlikely outlier. Please view future updates to the book for Author commentary of this purported AIDS cases.

The timing and location of this singular, disparate case places it outside the likelihood of HIV infection. Rather, if the clinical profile matches, this patient had one or another form of primary or secondary immunodeficiency.

Time Period:	1978 **X**
Number:	1 **X**
Location:	Sweden **X**
Methodology:	Clinical Report + Serum Analysis **X**

X - error – this purported AIDS case was listed by Vangroenweghe et al. (2000). However, Vangroenweghe et al. apparently made an error in citing this case

Description:

Child of parents treated in Sweden.

Reference: Clumeck N, Sonnet J, Taelman H, Cran S, Henrivaux P. Acquired immune deficiency syndrome in Belgium and its relation to Central Africa. Ann N Y Acad Sci. 1984;437:264-269 [204]

Comment:

Again, this listing in the original table seems to be an error.

The citation does not discuss any such patient.

Rather, Clumeck et al. describe the population of 40 Africans diagnosed in Belgium with AIDS between 1979 and 1983. (By comparison, the United States had reported 3000 AIDS cases by December 1983.[14]

The vast majority of AIDS cases diagnosed in Belgium were in African patients who lived in Belgium or who had traveled to Belgium seeking medical care.

All patients belonged to the upper socioeconomic class in their country.

Eighteen had been living in Belgium for less than 2 years at diagnosis, but had frequently returned to Africa for familial or business purposes. The other nineteen patients were living in Africa and came to Belgium because of unexplained fever, weight loss, or chronic diarrhea.

See *"The First African Patients – Diagnosed in Belgium"* for a discussion of this group and similar patient populations.

Time Period:	1979
Number:	---
Location:	Zaire
Methodology:	---

Description:

Epidemic of AIDS in Zaire

Reference: Vangroenweghe D, Afrique. Sese, Bruxelles BEE, c2000. Sida et sexualité en Afrique EPO; 2000 [496]

Comment:

This source article was not yet available at press time.

The time frame of this purported epidemic does allow the possibility of HIV being in Zaire.

By 1983, there were probably legitimate HIV/AIDS cases in Zaire, brought there by closeted gays from the Paris/Brussels epicenter. However, the presence of legitimate Zairean AIDS cases in 1979 would seem unlikely.

Time Period:	1980
Number:	3%
Location:	Léopoldville, Belgian Congo
Methodology:	Serum Analysis

Description:

Described only as "Zairean mothers in Léopoldville. Evidently, Desmyter (et al.?) conducted an HIV assay survey of some female population in Léopoldville, Zaire and found 3% seropositive.

Reference: Personal Communication, J. Desmyter

Comment: This finding is obviously based on a retrospective sero-survey conducted on stored serum or plasma. Such HIV assay outcomes are highly questionable. See *"HIV Antibody Testing – An Avalanche of False-Positives"*.

---

*now Kinshasa, Democratic Republic of the Congo

Time Period:	1980–1981
Number:	29
Location:	Burundi
Methodology:	Serum Analysis

Description:

A total of 658 serum samples collected in Burundi in 1980 – 1981 were tested for HIV antibodies. Twenty-nine (29) tested positive.

The overall prevalence was 4.4%. Among urban populations, prevalence was 8.08%. Among rural populations, prevalence was 2.82%.

The investigators also state what many other investigators before and after them stated the assay reactions might be the consequence of HIV or possibly some other retrovirus. The phrase "possibly some other retrovirus" is always lost in repetition.

Reference: Morvan J, Carteron B, Laroche R, Bouillet E, Teyssou R, Blanchard de Vaucouleurs F. [A sero-epidemiologic survey of HIV infection in Burundi between 1980 and 1981]. Bull Soc Pathol Exot Filiales. Jan 1989;82(1):130-140 [497]

Comment: Another sero-survey with questionable outcomes unless the urban cluster contained a gay cohort linked to Europe; but the high rural prevalence is nonsensical.

Time Period:	1982 – 1983
Number:	14
Location:	Paris (9)
	Brazzaville, Congo (4)
	Mali (1)
Methodology:	Clinical Report + Serum Analysis

Description:

In the abstract of the report (Katlama et al. 1984):

The first cases of AIDS in black Africans without any previously known risk factors were reported in March 1983. By April 1984, 24 such patients were reported in France: we studied 14 of them seen in Paris between 1982 and December 1983.

They all met the usual criteria for AIDS. Thirteen were adults (mean age: 31.5 years), one was the 12 months-old child of one of the female patients. Eight were males and six were females (sex ratio: 1.3). Nine of them were native from Zaire, four from Congo and one from Mali.

All were previously healthy. Opportunistic infections among these patients were: cytomegalovirus infection (6 cases), candidiasis (5 cases) and cryptococcosis (5 cases), *Pneumocystis carinii* pneumonia (PCP) (4 cases), neurotoxoplasmosis (3 cases), atypical mycobacteriosis (2 cases), cryptosporidiosis (2 cases).

The frequency of PCP was surprisingly low and that of cryptococcosis unusually high. Only one patient had Kaposi's sarcoma. Eight patients died (53 per cent), with a mean delay between onset of symptoms and death of 7 months. The mean follow-up in survivors is 10 months. Retrovirus serological studies were performed in our patients: HTLV-I-P antibodies could not be detected by radio-immunoassay. IgG antibodies to lymphadenopathy-associated-virus (LAV) were present in 10 of 13 adult patients.

Reference: Katlama C, Leport C, Matheron S, et al. Acquired immunodeficiency syndrome (AIDS) in Africans. Ann Soc Belg Med Trop. 1984;64(4):379-389 [498]

Comment:

This is the time frame and patient population corresponding to the first legitimate HIV/AIDS cases among Africans. See *"The First African Patients – Diagnosed in Belgium"* for a discussion of this group and similar patient populations.

Note the low PCP prevalence and high cryptococcosis prevalence. It is likely at least some of these patients represent legitimate HIV/AIDS cases, meaning they were likely closeted gay males who contract HIV in the discos or sexual venues frequently by elite, African, gay men in Europe. These males then passed HIV onto their wives and/or girlfriends, some of their children were born with HIV infection. This community also transmitted HIV to the elite, urban communities of Francophone Africa.

Pneumocystis carinii is reportedly ubiquitous worldwide, but its presence lacking among African patients purported to be infected with HIV, both inside and outside of Africa. The prevalence of cryptococcosis among this population is high; suggesting either that:

(1) *Cryptococcus neoformans* and/or *Cryptococcus gattii* are ambient in the African environment or endemic at non-pathogenic levels among Zaireans (either in the presence and/or in absence of HIV infection); therefore, cryptococcosis is a frequent early-onset opportunistic disease of HIV infection among Zairean populations and/or it is occurring more frequently among non-infected Zaireans as well;

(2) *Cryptococcus neoformans* and/or *Cryptococcus gattii* are endemic to Zaire and some recent event altered the equilibrium of the pathogen relative to the human population; giving rise to a rapid increase in the frequency of cryptococcosis infections either concurrent with and/or in absence of HIV infection.

Such recent events might include change in environmental and/or social factors that cause more frequent intersection of humans with transmission vectors; or perhaps a more virulent strain(s) and/or drug-resistant pathogenic strains came into existence around the same time as the advent of HIV/AIDS.

Other common pathogens are noted for having developed drug resistance, such as the etiological agents of tuberculosis and malaria in developing countries and Methicillin-resistant *Staphylococcus aureus* (MRSA) in the developed world.

Notes

Notes

Notes

www.ingramcontent.com/pod-product-compliance
Lightning Source LLC
Chambersburg PA
CBHW080330270326
41927CB00014B/3157